CHEM 122

Experiments in General Chemistry I Laboratory

August 2013

Pauline Hamilton, Ph.D.
Crystal Yau, Ph.D.
Kamruz Zaman, Ph.D.

Community College of Baltimore County
Catonsville•Dundalk•Essex

CHEM 122 Experiments in General Chemistry I Laboratory, August 2013
3rd Printing August 2014
Copyright © 2013 by the Department of Chemistry, Community College of Baltimore County –
Catonsville-Dundalk-Essex

Published by Academx Publishing Services, Inc., Bel Air, MD

All rights reserved. No part of this publication may be reproduced or transmitted in any form or by any means, electronic or mechanical, including photocopying, recording, or any information storage and retrieval system, without the written permission of the publisher.

Requests for permission to make copies of any part of the work should be mailed to:

Permissions Department
Academx Publishing Services, Inc.
P.O. Box 208
Sagamore Beach, MA 02562
http://www.academx.com

Printed in the United States of America

ISBN-10: 1-60036-632-5
ISBN-13: 978-1-60036-632-1

ACKNOWLEDGEMENTS

We wish to express our sincere appreciation and thanks to Doug Meise, Clarke Romans and Diane Winter for their suggestions and support in developing and refining the experiments in this manual. Furthermore, we wish to thank the CCBC Office of Instruction for funding the project that enabled us to present this common laboratory manual to our students.

Pauline Hamilton, Ph.D.
Crystal Yau, Ph.D.
Kamruz, Zaman, Ph.D.

Table of Contents

Chemistry Laboratory Safety Rules 1
Equipment in the Chemistry Laboratory 5
General Care of Glassware 6
How to Prepare for an Experiment 7
Laboratory Record Keeping: The Laboratory Notebook 9
Experiment 1: Uncertainty in Measurement 13
Experiment 2: Density 35
Experiment 3: Chromatography 43
Experiment 4: Composition of a Hydrate 55
Experiment 5: Synthesis of Tris(ethylenediamine)nickel(II) Chloride 63
Experiment 6: Chemical Reactions 71
Experiment 7: Acid-Base Titration: Standardization of a Solution 87
Experiment 8: Acid-Base Titration: Determination of Equivalent Weight of an Acid 99
Experiment 9: Calorimetry 105
Experiment 10: Atomic Spectroscopy 117
Experiment 11: Molecular Geometry and Molecular Polarity 131
Experiment 12: Molar Volume of an Ideal Gas 141

Review Information for CHEM 122 Lab Final Exam 153

APPENDICES:

Appendix 1: Review on Significant Figures A-1
Appendix 2: Preparation & Interpretation of Graphs A-5
Appendix 3: Graph Paper for Experiment 2 .... A-11
Appendix 4: Theoretical Wavelengths of the Line Spectrum of Hydrogen A-15
Appendix 5: Table of Electronegativity Values A-17
Appendix 6: The Formal Lab Report A-19
Appendix 7: Sample of a Formal Lab Report A-23
Appendix 8: Writing Exercises A-29
Appendix 9: How to Enter Subscripts & Superscripts A-33

CHEMISTRY LABORATORY SAFETY RULES

It is important to keep in mind that **accidents are not planned events**. They will happen when we least expect them. In order to ensure that a safe and orderly environment is maintained in the laboratory, each student must comply with the safety rules given below. Proper conduct is of utmost importance in the lab. Carelessness or refusal to adhere to instructions could lead to bodily harm and/or loss of property.

The list of safety rules below is not all-inclusive but it contains some of the most important requirements for your protection from bodily harm. Failure to follow these rules may result in immediate dismissal from the laboratory, and upon the discretion of the instructor or lab manager, could result in permanent dismissal.

1. **Safety goggles must be worn at all times.** Wearing safety goggles protect your eyes from chemical splashes and flying objects. As long as someone in the lab is working on an experiment, safety goggles must be worn. Even when you are finished with the experiment and merely doing calculations at your station, you should continue wearing your safety goggles. In the chemistry lab we require *splash-proof* goggles that fit snugly on the face, of the type similar to one shown in the figure on the right. If you wear prescription glasses or contact lenses, you must wear these goggles over your glasses or lenses.

   If you disobey this very important rule and a chemical or other irritant gets in your eyes, flush for at least 15 minutes with water at the eye bath, or the closest source of clean water. Force your eyes open during this period. If you wear contact lenses and you get chemicals in your eyes, it is best if you removed your lenses before flushing your eyes with water. This is because chemicals can get trapped between the lenses and your eyeball without being flushed out. To avoid potential eye injury, you will not be allowed to work in the lab without proper eye protection.

2. **Appropriate footwear must be worn.** Wearing proper shoes protect your feet from chemical splashes and shattered glass, and provide good traction as you move about the lab. Wear closed-toe shoes that completely cover your feet to protect you from spills and broken glass on the floor. Do not wear high heels or uppers made of cloth, woven leather strips or other woven material. Sandals and flip-flops are considered unsafe. You will not be allowed to work in the lab if you do not have appropriate footwear.

3. **Appropriate clothing must be worn.** Wearing proper clothing can help protect your skin in the event of accidental spills/splashes. You must wear clothes that protect areas such as midriffs and thighs that are more likely to come into contact with spilled or splashing chemicals. Midriffs must be completely covered. Short pants can be worn only if they extend to at least just above the knees. Do not wear clothing with loose-fitting sleeves or flowing scarves as they are a fire hazard and can catch on equipment and knock them over. Jewelry such as bracelets and rings (with stones) should be removed. They may catch on equipment and cause accidents and also increase the risk of prolonging contact of chemicals with your skin. Also, to prevent any damage to your

clothes, you might want to wear an old shirt over your street clothes or just wear old clothes to the lab. Some aprons and disposable gloves are available for your use if you so desire.

4. **Long hair must be tied back or worn in a bun.** Tying back long hair or wearing it in a bun prevents the hair from accidentally falling into an open flame and catching on fire or falling into your experiment when you lean over. If you have hair longer than shoulder-length, you are **required** to tie your hair back or wear it in a bun. We do not supply hair bands. Bring your own and keep them in your drawer.

5. **Food or drink must not be consumed in the lab.** Keeping food and drinks out of the laboratory minimizes the risk of your ingesting harmful chemicals. Food and drinks could become contaminated if they are placed where harmful chemicals are routinely handled. Plus, you could contaminate your experiments causing you to get poor results. Do not leave food or drinks on the lab bench. They <u>will</u> be confiscated and discarded.

6. **Your instructor must be informed about any medical condition that might require special consideration.** Avoiding a medical crisis is important. If you have any allergies, sensitivities, or medical conditions (especially including pregnancy) which might be aggravated by chemicals, be sure to inform your instructor **IN WRITING**, as soon as possible, so that any special arrangements which may be necessary can be made for your protection. If you are or think you may be pregnant, you should consult with your primary care physician as to whether the chemicals used in this course may be harmful to you or your fetus. If necessary we can supply your physician with a list of the chemicals you will be working with in the lab. Your physician may advise you to take extra precautions in certain experiments or possibly advise that you take the course at another time.

7. **All laboratory experiments must be supervised.** Working in the chemistry laboratory carries a certain amount of risks. To minimize the potential for harm, each experiment must be done under the supervision of trained personnel. Never work in the laboratory unless the instructor or lab manager is present. No unauthorized experiments are allowed.

8. **Students must be punctual.** Arriving on time ensures that you do not miss any important information. Your instructor will go over directions and point out safety precautions at the beginning of the lab. If you arrive late and miss these directions you may be prevented from performing the experiment.

9. **Follow directions carefully.** This includes directions in your lab manual, verbal directions given at the beginning of the lab and written directions that come with lab equipment. Always read the label on a bottle at least TWICE before dispensing chemicals from it.

10. **Memorize the location of all safety equipment** (fire extinguishers, eye bath, safety shower, first aid kit, etc.) in the lab *before* any emergencies happen. All labs have a

secondary emergency exit. Know where it is located. Report all cuts, burns, and accidents (no matter how minor) to the instructor and/or laboratory manager immediately. Also report all spills and breakage immediately.

11. **Keep your workspace orderly.** Do not place jackets or book bags under the hoods, on the lab benches, or on the floor next to your lab station. Keep **only** the necessary items (lab notebook, lab manual, pen) on the bench. The rest should be placed in an area of the room designated by your instructor. Keep your drawer closed when not in use. It is too easy for you and others in the lab to trip over an opened drawer. Keeping the drawer closed will also keep chemicals from falling in and contaminating the contents of your drawer.

12. **Do not chew on or carry your pens or pencils in your mouth.** You never know what toxic materials they may have picked up from the bench.

13. **Avoid touching your face, eyes or mouth** when working with lab reagents and chemicals. Use spatulas, spoonulas or Scoopulas to transfer solids from one container to another. Wash your hands immediately if they come in contact with any chemicals. At the end of the lab period, wash your hands thoroughly with soap and water.

14. **Never smell a chemical with your face close to it.** Waft vapors towards you with your hand.

15. **Never point the mouth of a test tube directly at anyone, including you.** This is especially important when the test tube is being heated. Direct it at the wall or ceiling of the lab. When heating a liquid, place the test tube holder high, near the mouth of the test tube so that the holder does not become hot. Hold the test tube at a slant and move the test tube in and out of the flame repeatedly to prevent hot spots from forming. Hot spots could cause the liquid to bump and eject from the test tube.

16. **Never add water <u>to any concentrated acid</u>**, especially sulfuric acid. The acid should be added <u>to the water</u>, slowly, so that any heat generated by the mixing process can be quickly absorbed and dissipated. It is also better that what might splatter out of the receiving container is just water, rather than concentrated acid.

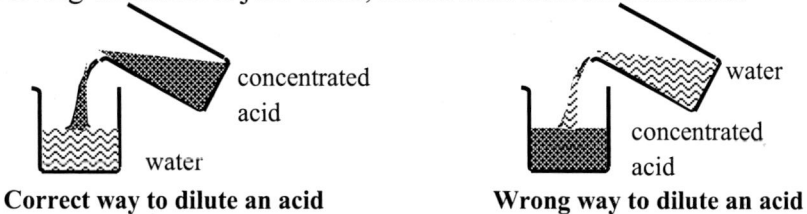

Correct way to dilute an acid   **Wrong way to dilute an acid**

17. **Don't walk away from a heat source** that is in use. Use mitts or tongs when handling hot materials. Glass, iron rings and wire gauze look cool long before they can be handled safely.

18. **Position power cords for electrical or computer apparatus so they are not near any water or heat source.** When returning them to their storage location, do not

allow the power cords to dangle down. Hotplates or heating mantles **must** be allowed to cool before returning them to the shelves.

19. **To avoid accidental contamination, never insert a spatula/spoonula/Scoopula or dropper into a stock bottle** (unless the bottle is a "dropper bottle" that comes equipped with its own dropper). **Never return used or unused chemicals to the stock bottle.** Follow directions on what to do with excess chemicals. Label all containers into which you put a chemical. After using any reagent bottle, replace the lid or stopper and wipe the outside of the bottle if any spills have occurred. Wipe up spills immediately.

20. **Dispose of each type of waste properly.** Instructions on disposal are specified on the blackboard and in the lab manual. Do not throw solid waste in the sink or water troughs. NEVER pour hazardous waste down the drain or into the trash cans. Do not pick up broken glass with your bare fingers. You must use a broom and a dustpan to collect the pieces. Broken glassware or porcelain should be placed in the special paper carton marked BROKEN GLASS ONLY. If you break a mercury thermometer do not attempt to pick up the mercury with your bare hands because mercury is hazardous to your health. Inform the instructor who will take care of the clean-up.

21. **Keep balances immaculately clean at all times.** Notify your instructor immediately of chemical spills. Zero the balances after use.

22. **Wash glassware after each use** and return it to its proper location (shelf or your drawer). Cleaning materials (soap, acetone, brushes, paper towels, etc.) are provided.

23. **Return all items to their proper location.** We will make every effort to provide designated storage location for all materials. If you cannot determine where something should be stored, consult your instructor or the lab technician. Do not leave any materials on the lab benches unless directed to do so.

EQUIPMENT IN THE CHEMISTRY LABORATORY

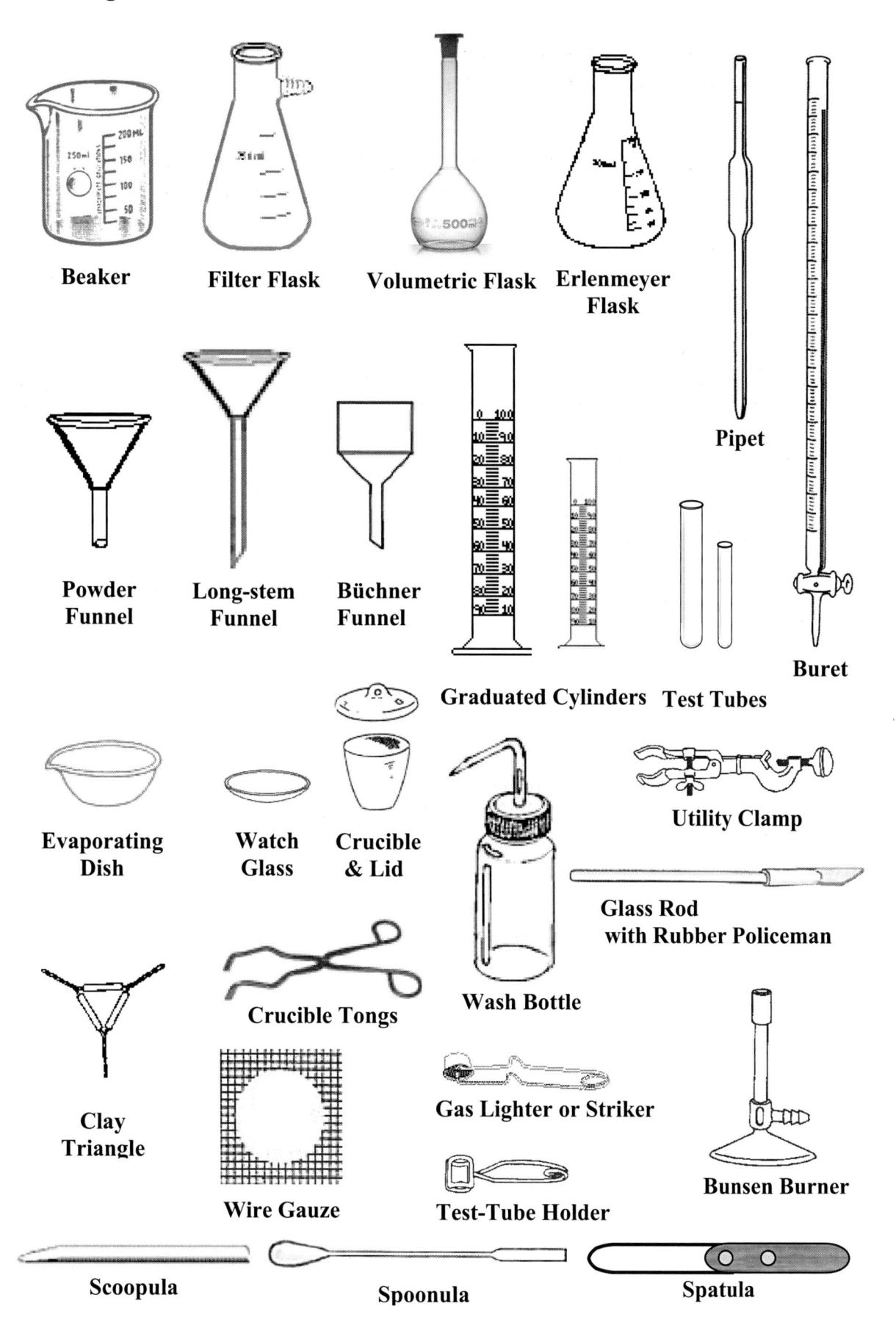

| Beaker | Filter Flask | Volumetric Flask | Erlenmeyer Flask | Pipet |

| Powder Funnel | Long-stem Funnel | Büchner Funnel | Graduated Cylinders | Test Tubes | Buret |

| Evaporating Dish | Watch Glass | Crucible & Lid | Wash Bottle | Utility Clamp |

Glass Rod with Rubber Policeman

| Clay Triangle | Crucible Tongs | | | |

| Wire Gauze | Gas Lighter or Striker | Test-Tube Holder | Bunsen Burner |

| Scoopula | Spoonula | Spatula |

GENERAL CARE OF GLASSWARE:

Equipment you get from the side shelf is not guaranteed to be clean. You have to use good judgment as to whether it needs to be cleaned before use. Do not wash everything indiscriminately as there are experiments where the equipment cannot be wet. Consult with your instructor when in doubt.

1. Follow instructions on how to dispose of the contents of your glassware. Instructions are provided on the blackboard in the lab or in the write-up of the experiment. Many of the chemicals cannot go down the drain because of toxicity to humans and/or to the environment. **FOLLOW INSTRUCTIONS!**

2. Remove any labels that are on the glassware. Labels are much harder to remove if left on the glassware for a prolonged period. In experiments where the glassware is heated and weighed, any labels that are not removed will burn partially during the process and affect your data.

3. Most of the chemicals you use in this course is water soluble. After the chemicals have been disposed of properly as instructed, the general rule in cleaning the glassware is to wash it with hot tap water and a **SMALL** amount of detergent. Scrub the inside and outside of the glassware with the brushes provided by the sink. The detergent is provided in a plastic bottle with a pump at the top. **DO NOT USE AN EXCESSIVE AMOUNT OF DETERGENT!** You will have a hard time getting rid of the detergent, which will act as a contaminant next time you use the equipment.

4. Next, rinse your glassware thoroughly with hot water to get rid of the detergent.

5. When you are satisfied that the glassware is clean, rinse it 2 or 3 times with small amounts of deionized water. This will get rid of the salts that are in the tap water. Do not waste deionized water by using excessive amounts. It is more efficient to rinse something several times with small amounts than to rinse it one time with a large amount.

6. Allow the water inside the glassware to drain out as much as you can, then wipe the outside with paper towels.

7. Glassware and all other equipment you obtain from the side shelf need to be returned to the side shelf for use by others in the next class. Please lock only your own equipment in your drawer. When in doubt, check with your instructor.

HOW TO PREPARE FOR AN EXPERIMENT

It is important to arrive at the laboratory well-prepared for the experiment of the day. As you will soon discover, reading an experimental write-up before performing the experiment is very different from reading a novel. You should expect to read the experiment at least three times: the **first time** to obtain the general idea of what the experiment is supposed to accomplish, the **second time** to obtain a general idea of the sequence of steps that you will be performing, and a **third time**, paying attention to precautionary details that are usually highlighted in bold, in italics, or underlined. In the second reading, you may find it useful to make a few sketches to help yourself understand the steps and setups in the procedure. In the third reading, you may wish to use a highlighter to further draw your attention to special instructions.

Just before you begin the actual execution of the procedure, you should reread the procedure, one paragraph at a time. You would not want to make the mistake of following the directions by reading one sentence at a time because the precautions may be at the end of the paragraph. Also, you should not rely solely on your memory of what your instructor had just demonstrated in the pre-lab discussion. Usually, your instructor is only showing portions of the procedure to demonstrate a specific technique. You must refer to the lab manual for the specific quantities and the exact sequence of steps in the procedure.

The Pre-Lab Exercise: For every experiment there will be some sort of Pre-Exercise to help you prepare for the experiment of the day. You will be asked to either submit the answers to these questions and/or take a quiz at the beginning of the lab period to show that you are properly prepared. These exercises or quizzes are worth about 10% of the course grade. There will be no make-up and late papers will not be accepted, **so do not arrive late.**

To prepare, study the entire experiment and pay particular attention to the concepts and definitions presented in the Introduction section, and to the safety precautions for the experiment. You are not expected to memorize details such as the size of a beaker and exact amount of a reactant but should know the general method used in the experiment.

The Preparation of the Lab Notebook: A specific type of lab notebook with carbonless duplicate pages is to be used. Your lab note book must be prepared before you arrive at the lab. More details on this preparation are provided in the next chapter. Your notebook will be checked at the beginning of each lab period and 5% will be deducted from the experiment grade if it is not properly prepared.

The Pre-Lab Discussion: There will be a pre-lab discussion the first 30 to 60 minutes of each lab period. It is crucial that you arrive on time. If you arrive more than 15 minutes late and miss a significant portion of this discussion you will not be allowed to do the experiment because of safety concerns and you will receive a zero for the experiment. In addition, the lab assignment for the previous week would be counted as late and incur a 10% late penalty.

During the pre-lab discussion your instructor will go over the pre-lab exercise and/or quiz, the finer points of the procedure, the safety precautions and some of the calculations of the experiment. Laboratory techniques will be demonstrated and the assignment will be specified.

The Lab Practical Final Exam: A Lab Practical Final Exam (worth about 15-20 % of the course grade) will be given at the end of the semester (see lab schedule provided). You need to prepare for this exam <u>throughout</u> the semester, rather than wait until the week before. Your letter grade will be significantly affected if you perform poorly in the final examination.

Students often have a false sense of security as to how well they are doing in this course. You may be getting good grades in your lab assignments because of the many sources of help available to you during the semester. We lead you through most of the calculations in the pre-lab discussion, help you with the calculations during the lab, and you have lab partners, tutors and friends to help you complete the assignment. All this is fine, and we expect you to get help whenever necessary to understand the material, but at the Final Exam, **<u>you will be on your own.</u>** It will be a **closed-book exam.** By the end of the semester, we expect you to have learned the material, and not to rely on notes, textbooks and the help of your instructor or lab partner.

Towards this end, you should go over the **<u>entire</u>** lab assignment for each experiment once it has been graded and returned to you. Get help from your instructor if there is anything you don't understand. This includes understanding the concepts, the calculations and laboratory techniques involved in each experiment. For your convenience a review chapter has been provided at the back of the lab manual.

IMPORTANT ADVICE: A common mistake is for students to wait until the day before the lab (or even just a couple of hours before the lab period) to complete the lab report that is due. Their time becomes totally occupied with the experiment of the **PREVIOUS** week rather than getting ready for the experiment of the **CURRENT** week. Do not underestimate the importance of being prepared for the experiment of the week. You might be able to get by with a fairly good grade during the semester by getting help from fellow students and your instructor, but if you don't really understand what you are doing you will be hit hard at the Lab Final. It is not uncommon for students to get a very poor grade on the Lab Final and end up with a lower grade for the course than they expected.

LABORATORY RECORD KEEPING: THE LAB NOTEBOOK

All students should purchase the chemistry lab notebook from the CCBC bookstore. This is a specific type of lab notebook which allows experimental data to be recorded in duplicate – the original always stay in the notebook and the carbon copy is submitted to the instructor at the end of every lab period. **No original pages are to be ever torn out of this notebook, no matter how "messy" they may be.**

> The notebook provides a special flap that is to be placed between the original page and the duplicate page to prevent writing from showing up on several subsequent pages. **Every time you make an entry in this notebook it is important to check to see that this flap is in place or else subsequent duplicate pages will become illegible. This is particularly important when you flip back and forth to enter data on different pages in your notebook.**

The lab notebook is where you keep a record of what you are doing in the lab. It is a place to record data and observations DURING THE EXPERIMENT. It is not meant to be a formal lab report (which is described in the appendix). Entries are expected to be complete and legible, but you are not graded on neatness. Its contents must be intelligible not only to you, but to others skilled in chemistry.

At the front of the notebook is the **Table of Contents**. You are to keep this up to date throughout the semester.

Before you come to the lab you are expected to have prepared your lab notebook for the experiment. This includes the following:

1. Each experiment must begin on a new page. At the top of each page fill out the information requested, such as the experiment number, experiment title, date, your name, your partner's name and anything else that your instructor may specify.

2. **Purpose:** (Write this and subsequent headings in your notebook): State in one or two complete sentences what you are trying to accomplish in the experiment; do not copy verbatim from the laboratory manual.

3. **Reference:** Give the reference to the procedure you will be following in the proper format. An example is shown below.

4. **Data & Observations:** Prepare this section as specified in the experiment write-up. If a sample of a data table is provided, copy it into the lab notebook. Be sure to allow enough room to enter your data and observations. Sometimes you have to design your own data table. You do *NOT* want to waste time preparing these tables while you are doing the experiment. You may be penalized or forbidden to perform the experiment if it becomes apparent that you have arrived unprepared. Your lab notebook will be inspected at the beginning of each lab.

LABORATORY RECORD KEEPING: THE LAB NOTEBOOK

Record all **observation** such as color, formation of gas, temperature changes, smells, etc. *at the time of the experiment.* Do not wait until the end of the lab period to record this. It is also the place where **raw data** are to be recorded, such as the exact mass or volume of reagents used. For example, the procedure may specify that "2 g" of a reagent is to be used. This means roughly 2 grams (not exactly 2.000 g) is to be measured out, but the exact mass that you took is to be recorded *precisely* (to the maximum number of significant figures that appears on the balance display, such as 2.008 g). The exact mass must be used in the calculations, not the approximate mass.

Data should be properly labeled so that not just you, but anyone reading your lab notebook (along with the procedure in the lab manual) would know exactly what you did. All numbers must be entered with the correct number of significant figures and proper units. If questions are posed within the procedure, you must answer them in the lab manual in sequence as you come across the questions.

At the end of the lab period, make sure you have collected all the pertinent data, signed and date in the appropriate location of the laboratory manual and then obtain your instructor's signature. Turn in all of the carbon copies for the experiment **before you leave.** You may not add extra data and observations once you have turned in your carbon copies.

Below is an example of how the lab notebook is to be prepared:

EXP. NUMBER	EXPERIMENT/SUBJECT	DATE
# 17	Determination of the Conc of an Aq Soln of HCl	2/3/06
NAME	LAB PARTNER	LOCKER/DESK NO.	COURSE & SECTION NO.
John Black	M. Smith	207	CHEM 122 – CM1

Purpose: The concentration of an unknown HCl solution is to be determined.

Reference: Hamilton, P., Yau, C. and Zaman, K., *CHEM 122 Experiments in General Chemistry I Laboratory,* Academx Publishing Services: Bel Air, MD, 2013; pp 138–147

Data & Observations:

HCl Unknown # = _____
Volume of HCl solution used in each titration = _____
Concentration of NaOH solution = _____

Trial #	# 1	# 2	#3
Final Buret Reading of NaOH			
Initial Buret Reading of NaOH			
Volume of NaOH soln used			

SIGNATURE	DATE	WITNESS/TA	DATE
	2/3/06		2/3/06

LABORATORY RECORD KEEPING: THE LAB NOTEBOOK

Use a **ruler** to draw in the lines to your data table.

A common mistake in preparing the data table is to take up too much space in the first column (stating what the data will be) and leaving too little space for the data itself. This is easily taken care of if you allot two or more rows for each entry, especially if you tend to write big.

AVOID TAKING TOO MUCH SPACE: (too little space for the data)

Trial Number	#1	#2	#3
Final Buret Reading of NaOH (mL)	23.76	23.87	23.26

IT IS MORE APPROPRIATE TO USE MORE THAN ONE LINE, IF NECESSARY (in order to allow more space for the data to be entered).

Trial Number	#1	#2	#3
Final Buret Reading of NaOH (mL)	23.76	23.87	23.26

Remember not to crowd your work. There are plenty of pages in your lab notebook. There is no reason for trying to fit everything on one page.

Recording Data:

Data should be recorded **in BLACK or BLUE INK, directly into the lab notebook**. They must **NEVER** be written on scratch paper to be copied into the notebook later. It is important for students to build the habit of recording data **immediately**, in a permanent location. Data, once entered, **may not be erased or obliterated.** It is to be a permanent record of what you observed. If you had to repeat the measurement and "change" the data, you may do so by neatly crossing out the previous entry and recording the new data directly next to or above it. The original entry **must still be legible**! This is how experiments in the professional scientific world are conducted. Erasing or obliterating an earlier entry are viewed as a sign of dishonesty and that attempts were made to hide something. You should try to be neat, but a certain amount of "messiness" is tolerated, as long as the entry is clearly legible.

<u>Acceptable Method of Correcting Data</u>
    Mass of Beaker + NaCl used = ~~2.177 g~~ 2.184 g

<u>Unacceptable Method of Correcting Data</u>
    Mass of Beaker + NaCl used = ~~2.177 g~~ 2.184 g } NO!
    Mass of Beaker + NaCl used = 2.1~~94~~ g

You should get into the habit of **immediately** checking your significant figures and recording the unit of any measurement, rather than returning to correct it later.

Calculations:

Simple calculations such as the subtraction of two numbers to obtain the mass of a sample should be performed **immediately** and entered in the lab notebook. You often need to know exactly how much you have measured out to see whether you have the correct amount.

Unless specified otherwise, other calculations are to be done on the "Calculations & Results" pages provided at the end of each experimental write-up in this lab manual. The pages are perforated so that you can detach and staple them to your report. **Please do not hand in pages with ragged edges**. The edges catch on other papers and make the handling of your papers an awkward and annoying task.

Unlike raw data (numbers that you record during the experiment, ones that do not require any calculations), which must be recorded in ink, calculations should be done in pencil. As a beginning chemistry student, you are likely to make mistakes in your calculations and in figuring out how many significant figures to use. Erasures of calculations and calculated results are acceptable. This is why you are requested to always have a pen (for the raw data) and pencil (for the calculations) with you. You might consider leaving them in your drawer.

How to show calculation setups: It is insufficient to just give an equation. You may start with an equation, but it must be followed with the actual numbers (with units and in the correct number of significant figures) you are going to use in your calculations.

Incomplete calculation setup: $d = \dfrac{m}{V}$

Acceptable calculation setup: $d = \dfrac{m}{V} = \dfrac{3.487 \text{ g}}{1.48 \text{ mL}} = 2.36 \text{ g/mL}$

Experiment 1:
UNCERTAINTY IN MEASUREMENT

Purpose: Determine uncertainty in measurements for various chemistry laboratory apparatus and take the uncertainty under consideration in the construction and analysis of a graph

Performance Goals:
- Determine the appropriate number of decimal places to use in recording measurements
- Utilize calibration scales to determine the accuracy and precision of a laboratory instrument or tool
- Apply statistical functions in the analysis of experimental data
- Construct a graph by hand and by the use of Excel based on data collected
- Analyze a graph to determine the accuracy and precision of the data collected

Introduction:

Laboratory work involves making reliable measurements. Reliability of a measurement depends primarily upon two factors: accuracy and precision. These two factors depend upon the quality of the measuring device, the procedure, and the technique of the operator.

The *accuracy* of a measurement is how close the measurement is to the correct or commonly accepted value. When we repeat the same measurement several times, the values obtained are not usually exactly the same. The *precision* is how close those measurements are to each other, that is, how reproducible the measurements are.

This is illustrated by the figure below showing the hits made by darts on a target. Good accuracy would be having a dart hitting bull's eye. Good precision would be hitting the same spot consistently, but not necessarily hitting bull's eye.

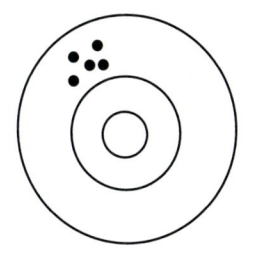

Poor accuracy
Good precision

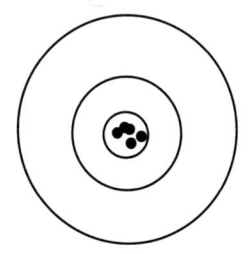

Good accuracy
Good precision

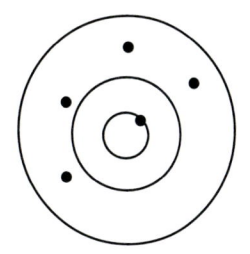

Poor accuracy
Poor precision

In this experiment we will begin by examining the graduations marked on various laboratory apparatus. The finer the graduation, the more precise is the measurement (the more reliable digits that can be recorded). **The general rule is to record to one-tenth of the smallest increment in the scale of the measuring device by estimating the last digit.** Thus, the last digit of any measurement has a degree of uncertainty.

For example, for Ruler I, shown below, the smallest increment is 0.1 cm and all readings should be recorded to one-tenth of 0.1 cm, which is 0.01 cm (2 decimal places). For Ruler II, however, the smallest increment is 1 cm, and all readings should be recorded to one-tenth of 1 cm, which is 0.1 cm (1 decimal place). You should follow this rule throughout the semester whenever you are making a measurement, unless the procedure specifically tells you otherwise.

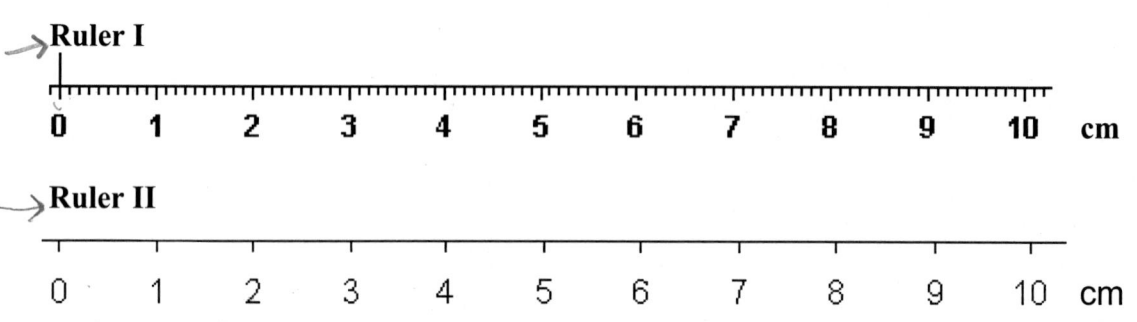

Ruler I

Ruler II

In the example below, the length of Object A, measured with Ruler I might be recorded as 8.14 cm, 8.15 cm or 8.16 cm. The last digit (the second decimal place) is estimated and is said to have "a degree of uncertainty." This is how all measurements should be recorded, and it is understood that when a number is reported, that the last digit has this *uncertainty in measurement*. The digits reported are known as *significant figures*. For a more extensive coverage on *significant figures*, refer to Appendix 1 and your chemistry textbook.

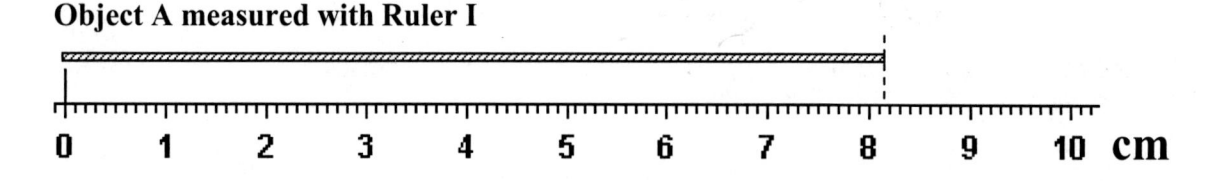

Object A measured with Ruler I

In the next example, the length of Object A, measured with Ruler II might be recorded as 8.1 cm or 8.2 cm. We can estimate only to the first decimal place because that already has uncertainty. We cannot estimate further into the second decimal place. Thus, it would be incorrect to record 8.10 cm or 8.20 cm.

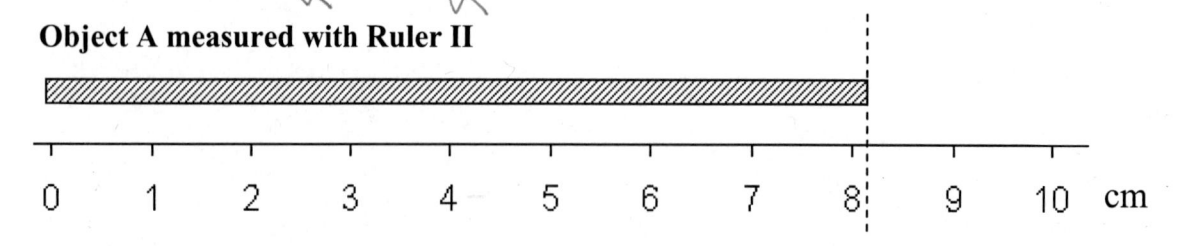

Object A measured with Ruler II

One of the errors most frequently made by beginning science students occurs when a measurement falls "exactly" on the line, as shown in figure below for Object B. Students often record the measurement as 5 cm, and if asked what the measurement is, they reply, "Exactly 5 cm." How "exact" is "5 cm"? It is not very exact at all! Remember that the last digit is assumed to be estimated. By recording only one digit such as "5 cm" you are implying that it could have been 4 cm, or it could have been 6 cm, but in actuality it could not have been. Instead it should be recorded as 5.00 cm.

EXPERIMENT 1: UNCERTAINTY IN MEASUREMENT   15

Object B measured with Ruler I
(This should not be recorded as 5 cm or 5.0 cm, but as 5.00 cm.)

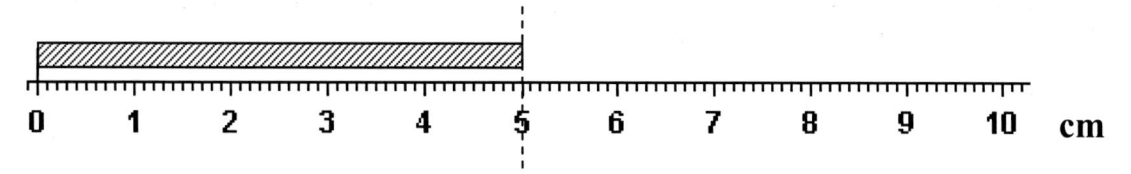

The measurement "5 cm" would be more appropriate for Ruler III shown below, where markings are only at 10-cm intervals (0, 10, 20, 30, 40…cm).

Object B measured with Ruler III *("5 cm" would imply a ruler of this sort was used.)*

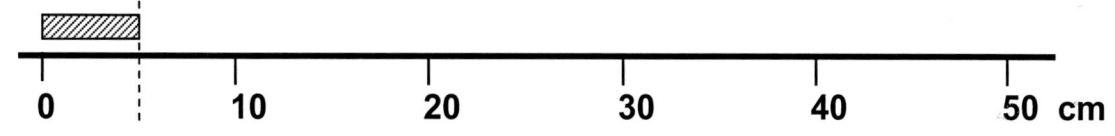

How do the graduations on a scale affect the precision of the measurement?

With a little bit of thought one should be able to conclude that the finer the graduations are on a scale, the more precise the measurements can be. Thus, Ruler I, graduated to 0.1 cm can pinpoint for us that Object A is around 8.12 cm; whereas, Ruler II, graduated to only 1 cm can only tell us it is around 8.1 cm. Use of Ruler I would give us more precise measurements.

When a measurement falls right "on the line" simply remember the rule of recording to one-tenth of the smallest increment in the measuring device.

What is the precision of a measuring device that has only one calibration mark?

One of the apparatus you will examine, in Part II of this experiment, is the volumetric pipet (or transfer pipet). Unlike the graduated cylinder it has only one calibrated mark. In this experiment you will work with a 10-mL pipet and determine how precisely one can measure 10 mL of water with this apparatus.

Graphical Analysis of Data:

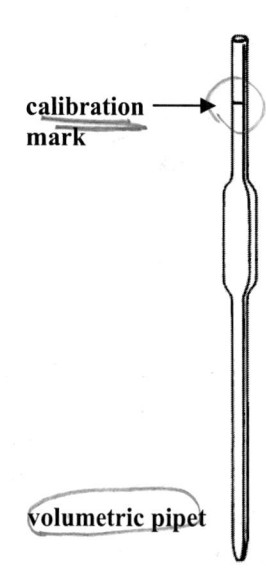

Analysis of data by constructing a graph is a way of finding the relationship between two variables. It also allows us to quickly see whether any part of the data collected is in error, and if so, whether the error is significant. Any piece of data that does not seem to fit in with the rest of the data is called an *outlier*. In Part III of this experiment you will heat up a sample of water and then record its temperature in °C and °F as it cools down. By constructing a graph of the temperature in °F versus the temperature in °C, you will obtain a mathematical relationship between the two temperature scale and at the same time be able to detect any significant outliers.

Error Analysis:

After an experiment the investigator is expected to reflect on what he or she did in the experiment and determine possible errors that might have occurred. Throughout the semester you will often be asked to provide possible sources of error. Some errors are inherent to the apparatus used (***Instrumental Error***). Some are due to the poor technique of the experimenter or failure of the experimenter to follow the procedure correctly (***Operator Error***). Some are due to the method of measurement or due to the assumptions made in the design of the experiment (***Method Error***) and cannot be avoided unless the design is revised. It is important to learn to recognize such unavoidable errors as it will teach you to design better experiments.

One of the post-lab questions in this experiment gives you examples of possible experimental errors related to this experiment. In future lab reports you may find it helpful to use these examples as a guideline for you to come up with sources of error on your own.

Equipment/Materials:

12"-Ruler (graduated in cm), 10-mL graduated cylinder, 50-mL graduated cylinder, 50-mL buret, 50-mL beaker, 250-mL beaker, 10-mL pipet, pipet bulb or pipet pump, digital thermometer, electronic balance, gas burner (Bunsen or Fisher), boiling chips

Procedure: *Work with one partner in gathering data, but do calculations individually. Remember that the raw data must be recorded directly in your lab notebook and not on scratch paper or in the lab manual. Before arriving at the lab, prepare data tables in your lab notebook similar to the ones shown below.*

Part I: Uncertainty in Measurement Using Graduated Apparatus

Examine each of the following measuring devices and record in your lab notebook, the size of its smallest increment, including the units. For example, in the figure for Ruler III on the previous page, the smallest increment 10 cm. One-tenth of 10 cm is 1 cm and therefore measurement should be recorded with no decimal places.

      i) Ruler (graduated in cm)
      ii) 10-mL Graduated Cylinder
      iii) 50-mL Graduated Cylinder
      iv) 50-mL Buret
      v) 250-mL Beaker

EXPERIMENT 1: UNCERTAINTY IN MEASUREMENT

Data Table for Part I: Uncertainty in Measurement Using Graduated Apparatus

Name of Apparatus	Size of Smallest Increment on Apparatus	Number of Decimal Places to be Recorded
Ruler (in cm)		
10-mL Graduated Cylinder		
50-mL Graduated Cylinder		
50-mL Buret		
250-mL Beaker		

Part II: Uncertainty in Measurement of a Volumetric Pipet

1. Record the mass of a clean and dry 50-mL beaker.
   Reminder: The balance should always be zeroed before weighing and you should always record all the digits that show on the display of the balance. Make a habit to use the same balance throughout this experiment to minimize instrumental error.

2. Place about 100 mL of deionized water at room temperature in a 250-mL beaker. Let it sit on the lab bench for at least 5 minutes and then record the temperature of the water in degree Celsius.

3. Obtain a 10-mL pipet. Record into your lab notebook, the *tolerance* marking on the pipet. This is usually in very fine print on the glass, usually as ± some numerical value). This tolerance represents the reliability of the pipet if it is used correctly by an experienced operator. It is what you, as a chemistry student, should strive for in developing your pipetting skills.

4. Remove the thermometer from the 250-mL beaker. Using the 10-mL pipet, deliver 10 mL of the deionized water into the pre-weighed 50-mL beaker. Before you proceed, read the instructions described below.

Proper Use of the Pipet and Pipet Pump
Never pipet by mouth in the chemistry laboratory!

The steps to transferring a desired volume of liquid by pipet varies somewhat depending on the type of pipet pump available for use. Determine which type of pipet pump is available to you and follow the procedure for your type of pipet pump.

Fast Release Pipet Pump with Side Release Lever

Fast Release Pipet Pump with Trigger Release Lever

Regular Pipet Pump

18 EXPERIMENT 1: UNCERTAINTY IN MEASUREMENT

Procedure for the Fast Release Pipet Pumps:

a) Examine the pipet pump. First ascertain that the bottom of the pipet pump is dry. It is difficult to operate a pipet pump that is wet. Also, it is likely that it will introduce contamination in your samples. If it is wet, consult with your instructor, who will either give you another one or help you dry it.
b) Examine the pipet. If it looks wet or dirty, consult with your instructor as to the procedure for cleaning it.
c) Attach the pipet securely to the pipet pump.
d) Hold the pipet pump with one hand and the pipet with the other, keeping it vertical. Immerse the tip of the pipet about 2 cm below the surface of the water in the 250-mL beaker.
e) Using the thumb wheel draw the water up into the pipet until the bottom of the meniscus is exactly on the calibration mark. *Do not allow the water to reach the top of the pipet and get into the pipet pump.*
f) Lift the pipet tip above the surface of the liquid. If there are any drops hanging outside the tip of the pipet, remove it by touching the tip to the inside of the 250-mL beaker.
g) Hold the pipet over the pre-weighed 50-mL beaker and press the release lever on the side (or the press the trigger). Allow the water to drain slowly into the beaker. If there are any drops hanging on the outside of the pipet tip, touch the tip to the 50-mL beaker. DO **NOT** BLOW OUT OR SHAKE OUT THE SMALL AMOUNT INSIDE THE PIPET TIP. The calibration mark takes into account that there will be a small amount remaining in the pipet tip.

Procedure for the Regular Pipet Pump:

a) Examine the pipet pump. First ascertain that the bottom of the pipet pump is dry. It is difficult to operate a pipet pump that is wet. Also, it is likely that it will introduce contamination in your samples. If it is wet, consult with your instructor; he/she will either give you another one or help you dry it.
b) Examine the pipet. If it looks wet or dirty, consult with your instructor as to the procedure for cleaning it.
c) Fit the top of the pipet to the bottom of the pipet pump *just* snug enough to effect a seal. Do not attempt to push it too far up. You must be able to remove the pipet easily at a later step. You should be holding the pipet pump with one hand and the pipet with the other hand. *You should not be dangling the pipet just by holding the pump. The pipet is **not** meant to be fastened that tightly to this type of pump.*
d) Immerse the tip of the pipet about 2 cm below the surface of the water in the 250-mL beaker. Applying a gentle pressure downward, pressing the pump onto the pipet. Use the thumb wheel to draw the water up into the pipet. You should see the water slowly rise up into the pipet. If you don't see the water rising, you do not

> have a tight enough seal to the pump. Re-position the pipet and pump and try again.
> e) When the level of the water has risen about 2 or 3 cm above the calibration mark, quickly remove the pump and place your index over the top of the pipet.
> f) Rock your index finger back and forth to find a position where the level of the water begins to slowly drop down towards the calibration mark. When the bottom of the meniscus reaches the calibration mark, press your index finger tightly down to prevent the water from draining any further. *If the meniscus should drop below the calibration mark, you must repeat Steps (e) and (f) until you have it just right.*
> g) While still holding your index finger tightly on the pipet, lift the pipet tip above the surface of the liquid. If there are any drops hanging outside the tip of the pipet, remove it by touching the tip to the inside of the 250-beaker.
> h) Hold the pipet over the pre-weighed 50-mL beaker and lift up your index finger. Allow the water to drain slowly into the beaker. If there are any drops hanging on the outside of the pipet tip, touch the tip to the 50-mL beaker. DO **NOT** BLOW OUT OR SHAKE OUT THE SMALL AMOUNT INSIDE THE PIPET TIP. The calibration mark takes into account that there will be a small amount remaining in the pipet tip.
>
> *Pipetting is a skill that takes time to learn. No one does it perfectly the first time.*

5. Record the mass of the 50-mL beaker containing the water. Calculate the mass of the water in the beaker. This is called *weighing by difference*.
6. Dry the 50-mL beaker thoroughly with paper towels (inside and outside).
7. Repeat the weighing and pipetting procedure (Steps 3–5) four more times. You need to weigh the empty beaker only once and assume it is the same for all five trials.
8. Look up the density of water in the CRC Handbook of Chemistry & Physics at the water temperature recorded.

   The following is a section from the density table from a CRC handbook available in the lab. It probably does not contain the section of the table that you need but serves to show you how to read the table correctly.

   The densities are given in 6 decimal places where the first 3 digits are the same for specified groups of densities. In the table shown below, all of the densities for 16–19 °C have in common, 0.998 as the first 3 digits and are not repeated in the table. When you look up the density for your particular temperature, do not forget to look up what the first 3 digits are for your temperature (may not be 0.998).

Absolute Density of Water
(density in grams per cubic centimeter)

Degrees	0	1	2	3	4	5	6	7	8	9
16	0.998943	926	910	893	877	860	843	826	809	792
17		774	757	739	722	704	686	668	650	632	613
18		595	576	558	539	520	501	482	463	444	424
19		405	385	365	345	325	305	285	265	244	224

For example, the density at 18.4 °C is 0.998520 g/cm^3. Remember that by definition 1 cm^3 = 1 mL exactly, thus it can be recorded as 0.998520 g/mL.

20   EXPERIMENT 1: UNCERTAINTY IN MEASUREMENT

→ Record the reference in the format commonly used for reference books in your lab notebook. *Below is an example of the format. Substitute in the appropriate edition and page number:* CRC Handbook of Chemistry and Physics, 47th ed. Weast, R.C., Ed.; CRC Press: Cleveland, OH, 1966-1967; p. F4.

Data for Part II: Uncertainty in Measurement with 10-mL Pipet

Temperature of Water: _____

Tolerance of the Pipet (printed on the pipet stem): _____

Trial #	Mass Empty Beaker	Mass Beaker + Water	Mass Water
1			
2			
3			
4			
5			

Density of Water (from Handbook) = _____

Reference: _____

Calculations:

On the Calculation & Results Pages, calculate the following:
1) Volume of water delivered by the pipet for each trial (calculated from the measured mass and the density of the water from the handbook)
2) Average volume of water delivered by the pipet

$$V_{average} = \frac{\sum V_i}{\text{\# of trials}} \quad \text{where } V_i = V \text{ of each trial}$$

3) Deviation of the volume of each trial from the average volume.
   Deviation = $|V_i - V_{Average}|$ *(Note the absolute value sign.)*
4) Average deviation, which is the average of the deviations of the five trials. (This is called the *absolute deviation.*)

   Absolute Deviation = Average Deviation

   $$= \frac{\sum \text{Deviations of each trial from } V_{average}}{\text{\# of trials}}$$

5) Percent average deviation (This is called the *relative average deviation or RAD.*)

   RAD = Relative Average Deviation = Percent Average Deviation

   $$= \left(\frac{\text{Average Deviation}}{V_{average}} \right) \times 100$$

Sample Calculations: A scientist wished to check the calibration of a 10-mL pipet which she intended to use for accurate delivery of 10-mL volumes in her research. She

EXPERIMENT 1: UNCERTAINTY IN MEASUREMENT

carefully pipetted five 10-mL samples of distilled water (at 22.1°C) into pre-weighed 50-mL beakers. The data she obtained is presented below:

Trial	Mass of Distilled Water (g)
1	9.9389
2	9.9669
3	9.9686
4	9.9767
5	9.9608

At this temperature (22.1 °C) the density of water is 0.997747 g/mL. Since volume is equal to the mass divided by the density, the following volumes can be calculated in the following manner:

$$9.9389 \text{ g} \left(\frac{1 \text{ mL}}{0.997747 \text{ g}} \right) = 9.96134 \text{ mL}$$

Trial	Volume Distilled Water (mL)
1	9.9613
2	9.9894
3	9.9911
4	9.9992
5	9.9833

Total Volume = 49.9243 mL
Average Volume = (49.9243 mL)/5 = 9.9849 mL
Deviations in Volume = | Volume of each trial – Average Volume |

Analysis of the data gave the following:

	Volume (mL)	Deviation (mL)
Higher	9.9992	0.0143
	9.9911	0.0062
Median	9.9894	0.0045
	9.9833	0.0016
Lower	9.9613	0.0236
Total	49.9243	0.0502
Average	9.9849	0.0100

Average Deviation = 0.0502 mL/5 = 0.0100 mL

When a set of measurements is listed in numerical order, the middle value is known as the *median*. From the deviation, we see that the first uncertain digit is in the hundredth place. Thus, the *tolerance* of the pipet is determined to be 9.98 ± 0.01 mL.

Relative Average Deviation = $\dfrac{0.0100 \text{ mL}}{9.9849 \text{ mL}}$ x 100 = 0.100 % deviation

Part III: Graphical Analysis of Data

In this part of the experiment you will collect a set of data to give you the relationship between the Fahrenheit and Celsius temperature scales. ***Study Appendix 2 to review what you should have already learned about the preparation and interpretation of graphs.***

Procedure: *Work with one partner to obtain the data, but construct a graph individually.*

1. Set up the apparatus as shown in the figure below. First set up the ring stand, iron ring, wire gauze and Bunsen or Fisher burner. Adjust the height of the iron ring and wire gauze so that there is about 6 centimeters between the top of the burner and the wire gauze.

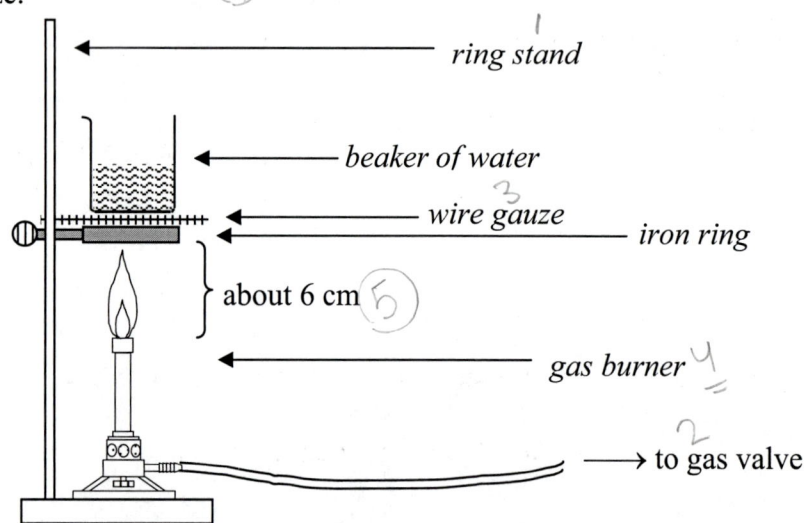

2. Place about 200 mL of hot tap water into a 400-mL beaker and add 2 or 3 boiling chips. The boiling chips have rough surfaces that enable the water to form bubbles easily as it boils. They are used as an aid to prevent the water from "bumping."

3. Light the burner with the striker. Be sure to adjust the burner to give a blue flame (rather than a yellow flame which produces less heat). This is done by opening up the chimney to allow more air to go in. (See Figure below.)

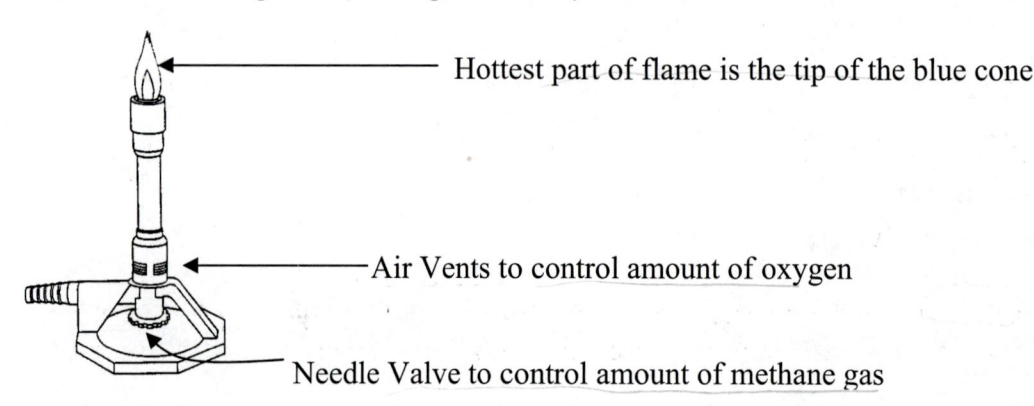

EXPERIMENT 1: UNCERTAINTY IN MEASUREMENT

4. Heat the water with the burner until the temperature has reached approximate 90 °C. In this experiment all temperatures must be recorded with a digital thermometer to 0.1 °C.

5. Meanwhile record the temperature, in °C and °F, of room temperature water that your instructor has set up in the front of the room. Always record all the digits that show on the temperature probe. Remember all data must be recorded directly in your notebook and not on scrap paper.

6. When your water has reached 90°C, one partner should call out the temperature readings in °C and °F in quick succession while the other partner records them. Temperature readings must be made with the probe held in the center of the water sample.

7. Turn off the burner and allow the water to cool. When the temperature has dropped approximately 2 °C, quickly record the temperature in both °C and in °F. Continue doing this until the temperature has reached 70.0 °C.
   Note: It is __not__ essential to be recording exactly at 90.0 °C, 88.0 °C, 86.0 °C, etc. For example, it is perfectly acceptable to record at 90.1 °C, 88.2 °C, and 86.3 °C (as long as they are approximately 2 degrees apart).

8. When you are finished, return all equipment to its proper place. Take extra precaution that your boiling chips do not get into the sink or down the drain. The temperature probes <u>MUST BE TURNED OFF</u> before being returned to the side shelf.

Before arriving at the lab, copy the data table below into your lab notebook.

DO NOT RECORD DATA ON THIS PAGE. Record directly in your lab notebook.

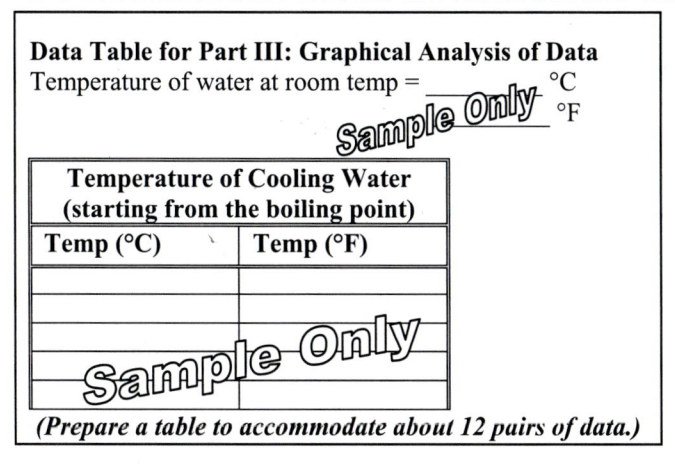

Data Table for Part III: Graphical Analysis of Data
Temperature of water at room temp = _____ °C
_____ °F

Temperature of Cooling Water (starting from the boiling point)	
Temp (°C)	Temp (°F)

(Prepare a table to accommodate about 12 pairs of data.)

Plotting the Graph Manually:
Read the directions below carefully before you begin plotting!
You may discuss with each other what to do but **each partner must do his/her own graph**. *Your slope and intercept are not expected to be exactly the same as those of your partner.*

1. **Use the blank graph paper provided in Appendix 3.** In the interest of time, you do not have to plot <u>all</u> the points for this graph. Select 5 points from your data, spread out fairly

24   EXPERIMENT 1: UNCERTAINTY IN MEASUREMENT

evenly between 90 °C and 70 °C. **Indicate clearly** in the data table in your lab notebook with asterisks to show the points you have selected for your graph. Using the graph paper provided, plot Temp (°F) vs. Temp (°C) for the temperature obtained as the water cooled to 70.0 °C. By convention this means Temp (°F) goes on the y-axis (the vertical axis) and the Temp (°C) goes on the x-axis (the horizontal axis). Each point should be plotted **with a sharp pencil**, by drawing an X such that the intersection of the X indicates the position of the point precisely. **Include the temperatures of the water at room temperature**. You will be plotting 6 points in total: 5 selected points and the point at room temperature.

2. Selection of a scale: Reread Appendix 2 on the proper selection of a scale. You need a scale that is not awkward to read and points that are not bunched up in a small area of the graph. (In particular, do not pick awkward scales such as using 6 squares for 5 degrees!) Since you will need to read the y-intercept on the graph, your x-scale should extend from zero to beyond the highest temperature you recorded in °C. For the y-scale, start with 20 °F and go past your highest reading in °F.

3. Your graph should indicate a linear relationship. Using a ruler and a sharp pencil, draw **one** best straight line through your points. **Do NOT draw several lines connecting dot-to-dot!** Your line may not necessarily pass through every single point. It should pass through most of the points, and if there are points that do not fall on the line, there should be as many points above as below the line.

4. Label the graph with a title at the top of the page. Next label each axis with a title, including the units used. At the top right hand corner, write your full name. When you are finished, show your graph to your partner and discuss with each other what improvements might be made. Before you proceed, show your graph to your instructor. Upon approval your graph will be initialed. Your graph is to be turned in at the next lab period.

Obtaining Information from the Graph:

$$°F = °C \times \frac{9}{5} + 32$$
$$°C = (°F - 32) \times 5/9$$

5. Interpolation: Using your graph, determine what the temperature in °C corresponds to 181.8 °F and record this on the Calculations & Results Page. Next, record the temperature in °F corresponding to 65.3 °C. Interpolation refers to making a prediction of values between two known values. If either of the two temperatures happens to be one of your data points, select another temperature to illustrate "interpolation."

6. Determine the slope of this line by choosing two points that lie <u>on the line</u> and are easily read. Do not use your own data points. Record the coordinates used and the calculated slope on the Calculations & Results Page. Note that since the temperature probe allows you to read to one decimal place, the coordinates should also be read to no more than one decimal place only. **Be sure to include units for the coordinates and for the slope**.

7. Read the y-intercept from the graph and record it on the Calculations & Results Page. This is done by extrapolating the line to x = zero. Review Appendix 2 if you are not sure

$$\text{slope} = m = \frac{y_2 - y_1}{x_2 - x_1} = \frac{-3 - 4}{5 - 2} = \frac{-7}{3} \quad (2, 4)$$
$$(5, -3)$$

what the *y*-intercept is. Before you record it, think about how many significant figures are appropriate. (Hint: How precise are the y-values you measured? The *y*-intercept cannot be more precise than your measurements.)

8. Deriving an Equation for the Graph: The equation of a straight line is commonly given as $y = mx + b$ where m is the slope and b is the *y*-intercept. Note that the slope is the coefficient of *x* in this equation. Using the slope and *y*-intercept you determined in Step 7 above write the equation for your graph on the Calculations & Results Page.

   Use the letter F to represent *y* and C for *x*. Do not confuse F with °F. F is a letter to represent various values of the temperature in degree Fahrenheit, similar to how one might use M for mass or V for volume. The symbol °F, however, is a unit for the temperature measured on the Fahrenheit scale. (For example, we can write F = 32 °F when water freezes.) The same goes for C and °C.

Plotting the Graph with the Use of Excel 2010: The objective is to create a computer-generated graph to obtain the equation of the line of best fit, which will give us the information we need. If lab tops are provided in the lab, work together with your partner on this graph and print out two copies, one for each of you. You should plan on finishing this graph in the lab before you leave.

If lab tops are not available in the lab, complete this exercise at home or at a computer lab on campus. Note: All CCBC college computers are using Excel 2010. If you have to create the graph at home on an older version of Excel these instructions will not be apply.

1. For this graph, do **not** include the temperature of the water at room temperature. You will be plotting **all** the points gathered from the water as it cooled down (not just 5).
2. In the A1 cell (1st row, column A), type "T(deg C)" and in the B1 cell, type "T(deg F)".
3. Enter your Celsius temperatures in column A, and the corresponding Fahrenheit temperatures in column B. (Enter only the numbers. Do not include units.)
4. In the D1 cell, type "CHEM 122 – Sec (*enter your sec #*) – Expt #1 Temperature Scales – (*enter your full name and your partner's full name*)."
5. Highlight both the A and B columns, from Row 1 down to the last row that contains an entry.
6. Click on the **Insert** tab, on **Scatter**, and then on **Scatter with only markers.**

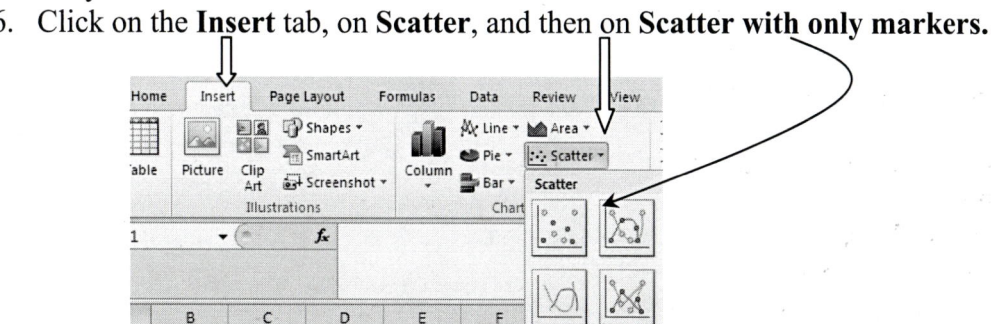

26   EXPERIMENT 1: UNCERTAINTY IN MEASUREMENT

7. You will be printing the graph embedded within the spreadsheet with both in Landscape Mode. Place your cursor anywhere on the spreadsheet (not on the graph) and click on **Page Layout** tab, **Orientation** and select **Landscape**. The dotted lines that appear indicate the size of the page you will be printing. Click on the frame of the graph and then resize the graph so that it is as large as possible without letting it spill beyond the dotted lines. *A common error is to set the graph instead of the spreadsheet on Landscape Mode.*

The graph you have created needs adjustments to the title, and to the scales because the default is to include the (0, 0) causing the points to be all bunched up in a small area. The following steps will allow you to format the graph to give titles to the graph and to each axis. You will also format the scales on the two axes as necessary to give a more acceptable graph.

*At this point you will see that a new tab (**Chart Tools**) has appeared. This tab appears only when a chart (i.e. graph) has been selected (highlighted). Directly below the **Chart Tools** tab are three sub-tabs: **Design**, **Layout**, and **Format**. The instructions below will focus on the most direct route to creating the graph we want and obtaining the equation of the line of best fit.*

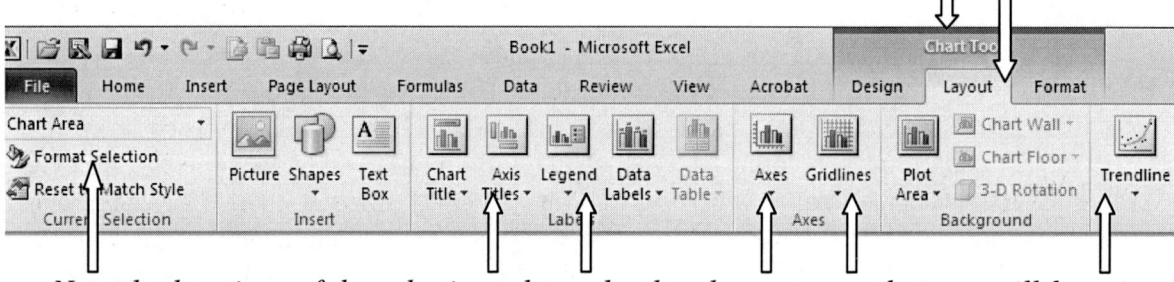

*Note the locations of the selections shown by the above arrows that you will be using under the **Chart Tools, Layout** in the next few steps.*

8. Highlight the title and replace it with a better title that must include your name and that of your partner's: "Temp in deg F vs. Temp in deg C - John Doe and Jane Smith".

9. Click on **Chart Tools (1), Layout (2), Axis Titles (3),** then select **Primary Horizontal Axis Title,** and **Title Below Axis.** Type in a title for the x-axis (Temp in deg C) and press **ENTER**.

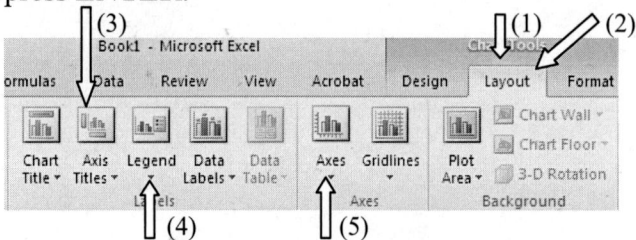

10. Click on **Axis Titles (3)**, then select **Primary Vertical Axis Title**, and **Rotated Title**. Type in a title for the *y*-axis (Temp in deg F) and press **ENTER**.

11. Click on **Legend (4)** and select **None.**

12. Click on **Axes (5)**, select **Primary Horizontal Axis** and then on **More Primary Horizontal Axis Options**. *Since the lowest x-value is probably slightly above 70°C, there is no reason why the minimum value has to be zero. This is where you are going to fix the problem of having all the points bunched up in a small area. You do not want a graph with large blank spaces because you would minimize the space for the section of the graph that you are interested in.* Select **Fixed** to allow you to change the default settings.

Set the minimum *x*-value at a number lower than your lowest x-value. (If your lowest value is something like 70.4 °C, set it at 70.) Similarly the maximums should be larger than the highest *x*-value you have recorded. (If your highest value is 99.8 °C, set it at 100.)

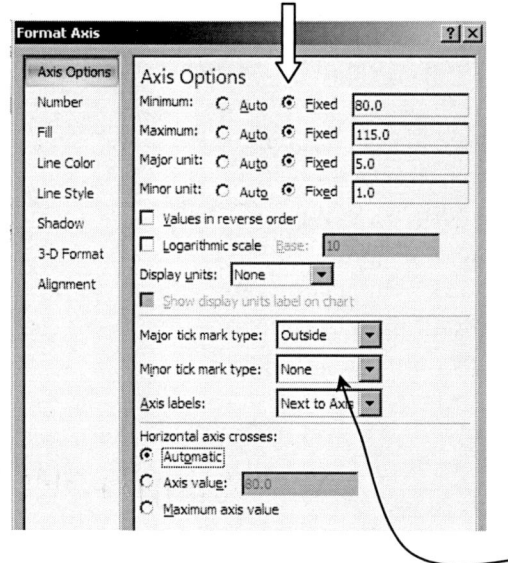

13. Adjust your **Major Unit** and **Minor Unit** so that the *x*-scale has reasonable looking numbers. Try **Major Unit** = 5 and **Minor Unit** = 1. Click on **Fixed** to allow you to enter these numbers.

    Major Unit controls which numbers are to show on the scale. *Major Unit of 5 means a number will appear every 5°C.* **Minor Unit** controls how far apart the minor tick marks are to be. *Minor Unit of 1 means the minor tick marks are 1°C apart.* This will be apparent only if the **Minor tick mark type** is not set at "**None.**" Go down to **Minor tick mark type** and change "**None**" to "**Outside.**" When you are finished, click on **Close**.

14. Adjust the scale for the *y*-axis in a similar manner. Click on **Axes**, select **Primary Vertical Axis** and then on **More Primary Vertical Axis Options**:
    In order to include all your data points in your graph, remember your minimum y-value cannot be larger than your smallest temperature in °F and your maximum cannot be smaller than your largest temperature in °F.
    Suggested settings (which may not work for you, depending on your data):
    Minimum = 150; Maximum = 220; Major unit = 10; Minor unit = 1 and change **Minor tick mark type = Outside**, then click on **Close**.

15. Click on **Gridlines**, select **Primary Horizontal Gridlines** and <u>Major</u> **Gridlines**.

16. Click on **Gridlines**, select **Primary Vertical Gridlines** and <u>Major & Minor</u> **Gridlines**.

Before proceeding, examine your graph. If there are any "outliers" first check to make sure it is not simply due to an error in entering the data. If it is due to some other error, consult with your instructor, who will decide whether you should discard that particular point.

28   EXPERIMENT 1: UNCERTAINTY IN MEASUREMENT

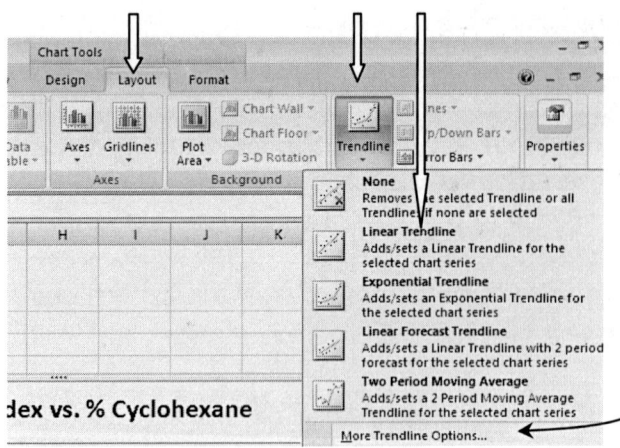

17. Next add the line of best fit by clicking **Layout, Trendline,** and selecting **Linear Trendline**.

18. To obtain the equation for the trendline, click again on **Layout, Trendline,** and select **More Trendline Options** at the very bottom. The **Format Trendline** window will then appear.

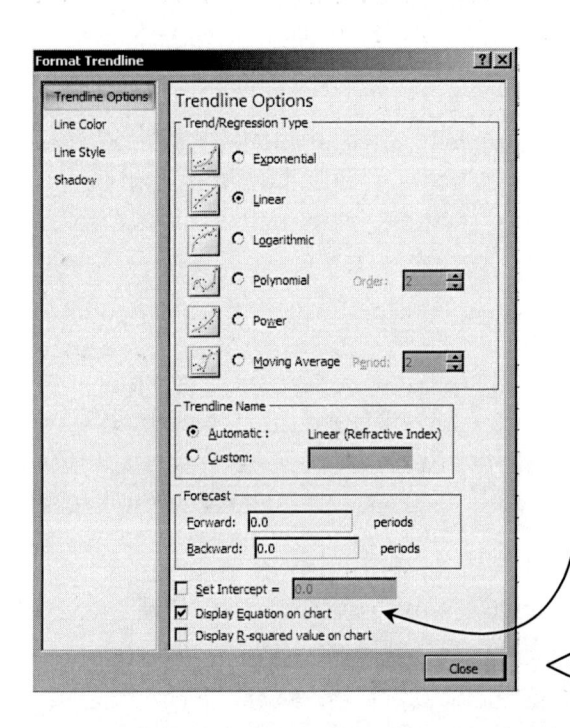

19. Place a check mark at **Display Equation on chart,** and click on **Close.**

20. If necessary, move the equation to a position where it can be read easily (such as just below the title). This is done by clicking on the equation once and then dragging it to the desired position.

$y = 1.4111x + 71.07$

Close

21. Highlight the area that contains your data table and the entire graph and click on **Page Layout, Print Area,** and select **Set Print Area**. *(If you have trouble highlighting the entire area with the touch pad, try this: Click once in the **A1** cell. Hold down **Shift** and use the right (→) and down arrows (↓) to highlight the area you want to print.)*

22. Go over this **CHECK LIST** before printing your graph:
    A) Spreadsheet: Columns must have proper headings and title must be at the top indicating course, section, experiment # and experiment title, your name and your partner's name.
    B) Graph must have title at the top and axes must be properly labeled.
    C) Trendline must be clearly shown at the top.
    D) Spreadsheet must be in Landscape Mode. *DOUBLE CHECK!*
    E) Graph must be enlarged as much as possible but fit within the page.

F) DOUBLE CHECK THAT THE CURSOR IS ON THE SPREAD SHEET JUST BEFORE YOU PRINT (to ensure graph is *embedded* in the spreadsheet.)
23. Click on **File** at the top left corner, select **Print**, and click on **Print Preview** to double check that everything fits on the page. If no adjustments are necessary, click on **Print**.

Calculations Based On the Excel-Generated Equation:
24. Copy the equation of the trendline (showing as is, with all digits) onto the Calculations & Results Page.
25. From the equation, copy the slope and the y-intercept onto your Calculations & Results Page. Re-write the equation by replacing x and y in the equation with C (for temp in °C) and F (for temp in °F).
    Note: C in this equation is a letter representing the temperature in °C, similar to using D for density. A common student misconception is to think C is the unit. In the same way, F is not a unit but the letter to represent the temperature in °F.
26. Using this new equation (written in terms of C and F) calculate the temperature in °C for 181.8 °F. Round it to 1 decimal place. Show your calculations in your Calculations & Results page.
27. Using the equation calculate the temperature in °F for 65.3 °C. Round it to 1 decimal place. Show your calculations in your Calculations & Results Page.
28. Are there any outliers in your data? Write your answer on the Calculations & Results page.

short-lab

Pre-Lab Exercise:
Before coming to the lab, be sure to study Appendix 2 on Preparation and Interpretation of Graphs. Your instructor will indicate whether you are to submit answers to the following questions at the beginning of the pre-lab discussion or you will take a pre-lab quiz with questions similar to the following.
1. If the smallest increment on a ruler is 0.01cm, to how many decimal places should measurements be recorded?

2. Name two major points to watch out for when selecting a scale for the graph other than making sure none of the data points are excluded.

3. If the equation of a line is given as $y = 7.8 x + 1.9$, what is the slope and y-intercept of this line?

4. If the y-intercept of a line is 2.31 and the slope is 0.825, what is the equation of this line?

5. The following are the coordinates of two points that lie on a line. (*s* stands for seconds.)
   (15.8 s, 78.3 cm) and (29.1 s, 94.2 cm)
   What is the slope of the line? Show your calculation setup. Include units in your setup and in your answer.

EXPERIMENT 1: UNCERTAINTY IN MEASUREMENT

Due tomorrow

Post-Lab Questions: H.W — Typed + Calculation + results + Prepare for Lab 2
Answers must be typed and in full sentences. Due Next tuesday!

1. How does your experimental tolerance of the 10-mL pipet compare to the tolerance stated on the pipet? If they are significantly different, give reasons why they are different. If they are not significantly different, what conclusions can you draw?

2. When you use this 10-mL pipet to deliver a liquid into a container, would you say you delivered 10 mL, 10.0 mL, 10.00 mL or 10.000 mL? Explain how you reached your answer.

3. Based on your graphs, were the temperature data you collected accurate? Explain how you reached your answer.

4. Based on your graph, were the temperature data you collected precise? Explain how you reached your answer.

5. In Part III, would you have arrived at the same equation if you had used a different liquid than water?

6. Give the location of each marker to the correct significant figure.

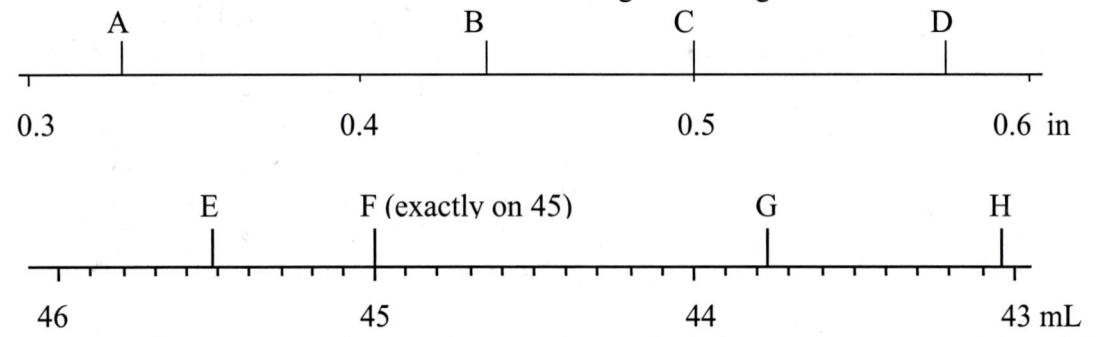

   A   B   C   D
   | | | |
   0.3   0.4   0.5   0.6 in

        E   F (exactly on 45)   G   H
        | | | |
       46   45   44   43 mL

7. You are told that a certain measurement has a deviation of 0.1 mL. Is this a large deviation? What if you were told that a measurement has a 0.1% deviation? Is that a large deviation? Explain your answers.

8. A student performing Part II neglected to dry the beaker thoroughly before the third trial. How would that affect the volume of water he reports for the third trial? Explain. Would this be Instrumental Error, Operator Error or Method Error?

Experiment 2: DENSITY

Purpose: Determine density of objects of different shapes using different types of laboratory tools

Performance goals:
- Determine the density of a regularly shaped object
- Determine the density of an irregularly shaped object
- Determine the density of a liquid
- Calculate percent error

Introduction:

Density is a physical property of matter, each element and compound has a unique density associated with it. Density can be described in a qualitative manner as the measure of the relative "heaviness" of objects with a constant volume. Density may also refer to how closely "packed" or "crowded" the material appears to be at the atomic or particulate level. For example, a rock is more dense than a crumpled piece of paper of the same size, or Styrofoam cup is less dense than a ceramic cup.

The density of an object is often compared to the density of water. Does an object float or sink in water? This is a question which is often asked. The answer will depend on the density of the object. For instance, a piece of wood floats on water because it is less dense than water while a rock sinks because it is more dense than water. Similarly, the oil in a salad dressing floats on the vinegar-water mixture because it is less dense, while the solids sink to the bottom because they are more dense. And, consider what happens when an oil tanker leaks on the ocean. The oil floats on the water since it is less dense, and this provides some opportunity to clean up the oil spills by skimming the oil from the surface of the water.

Density is an intensive property of matter; it does not depend on the amount of the matter. For example, an aluminum block and a piece of aluminum sheet have the same density if both of them come from the same source.

The scientific definition of *density* is mass per unit volume. $$\text{Density} = \frac{\text{mass}}{\text{volume}}$$

In order to determine the density of an object, it is necessary to know the mass and the volume of the substance. Usually the volume of a solid or liquid is expressed in milliliters (mL) or cubic centimeters (cm^3 or cc).

1 cm^3 is defined to be exactly 1 mL.

Thus, the density of a solid or liquid is expressed as grams per milliliters (g/mL or $g\ mL^{-1}$) or grams per cubic centimeters (g/cm^3 or $g\ cm^{-3}$).
For a gas, which is much less dense, the volume is usually in liters and its density is expressed as g/L or $g\ L^{-1}$.

35

EXPERIMENT 2: DENSITY

The volume of a regularly shaped solid object (such as sphere, cylinder, or cube) can be found by measuring pertinent dimensions and using the appropriate geometric formula to calculate the volume. The volume of irregularly shaped solids is measured by applying Archimedes' Principle which states that *the volume of a sample of a substance is equal to the volume of the water it displaces*. The solid is placed in a liquid which is less dense than the solid and in which it does not dissolve. When the solid sinks in the liquid it displaces or pushes aside a volume of liquid. This volume of displaced liquid can be measured and is equal to the volume of the solid. Table 2.1 lists densities of common elements and compounds at 25 °C.

Table 2.1: Densities of Common Substances at 25 °C

Substance	Density (g/mL)
Pine wood	0.35 – 0.50
Water	0.997
Salt, NaCl	2.16
Aluminum, Al	2.70
Iron, Fe	7.80
Mercury, Hg	13.5
Gold, Au	19.30

Equipment/Materials:

Wooden block, metal pellets, ruler (graduated in cm), analytical balance, 25-mL graduated cylinder, 10-mL graduated cylinder, water, unknown liquid

Procedure: *Prepare data tables as shown below for Parts I, II, and III in your lab notebook before arriving at the lab. Be sure to use the same measuring device for repeated measurements of the same substance.*

Part I: Density of a Regularly-Shaped Object

1. Take a wooden block and accurately measure its length, width, and height using a centimeter ruler. Record to 0.01 cm.
2. Use the formula, $v = l \times w \times h$ to find the volume of the wooden block. (v = volume, l = length, w = width, and h = height)
3. Record the mass of the wooden block.
4. Calculate the density of the wooden block to the correct number of significant figures on the Calculations & Results page.
5. Repeat steps (1)–(4) with the same wooden block.
6. Determine the average density of the wooden block on the Calculations & Results page.

EXPERIMENT 2: DENSITY

Part II: Density of an Irregularly-Shaped Object

1. Obtain a sample of metal pellets and record the unknown # of your sample in your laboratory note book.
2. Tare a weighing boat to zero and weigh out about 20 g of metal pellets. Record the exact mass of the metal pellets in your notebook.
3. Take a 10-mL graduated cylinder and pour about 5 mL of regular tap water into it. Record the volume accurately to 0.01 mL.
4. Slowly slide the metal pellets (from step 2) into the water. Tap the side of the graduated cylinder to dislodge any trapped air bubbles. Wait for a minute and record the water level accurately to 0.01 mL.
5. Calculate the volume of the metal pellets by difference.
6. Calculate the density of the metal pellets to the correct number of significant figures on the Calculations & Results page.
7. Thoroughly dry the metal pellets between paper towels and repeat steps (2)–(6). If the mass of the pellets is significantly heavier than the mass in Trial 1, your sample is not dry enough. Repeat the drying process and try again.
8. Determine the average density of the metal.
9. Get the true value of your metal sample from the instructor and calculate percent error.

Calculation of Error: *Error*, sometimes referred to as *absolute error*, indicates how accurate a measurement is; how far the measured value is from the true value (or commonly accepted value).

Error = | experimental value – true value | Equation 1
Error =   experimental value – true value   Equation 2

The absolute value sign in Equation 1 means the error thus calculated will always be a positive number. It indicates how far the experimental value is from the true value without indicating whether it is too high or too low.

Error calculated using Equation 2 will give a positive number if the measured value is higher than it should be, and a negative number if it is lower than it should be. Thus the algebraic sign of the error provides further information on the nature of the error.

Your instructor will specify which equation you are to use.

Calculation of Percent Error: *Percent error* is sometimes referred to as *relative error*. It indicates how far the measured value is from the true value on a percent basis.

$$\% \text{ Error} = \frac{\text{error}}{\text{true value}} \times 100 \quad \text{Equation 3}$$

It has the algebraic sign of *error*. If Equation 1 were used to calculate the error, % error would always be positive. If Equation 2 were used, the algebraic sign would be positive if the measured value is too high, and negative if it is too low.

Often it is difficult to judge how significant an error (absolute error) is because it would depend on the magnitude of the measurement. For example, an error of 1 g is large for an object that weighs 10 g, but it is small for an object that weighs 1000 g. In contrast, percent error (relative error) gives us perspective as it is always based on 100 regardless of the magnitude of the measurement.

EXPERIMENT 2: DENSITY

Part III: Density of an Unknown Liquid

1. Obtain a sample of unknown liquid and record the code # of the sample in your lab notebook.
2. If your 10-mL graduated cylinder is wet, dry it thoroughly with a piece of paper towel and record its weight.
3. With the use of a dropper, fill the graduated cylinder **exactly** to the 10-mL mark with the unknown liquid and weigh it again. Record the volume to the correct sig. fig.
4. Return the sample to its original container in the designated location.
5. Calculate the density of the unknown liquid to the correct number of significant figures on the Calculations & Results page.
6. Get the true density of your unknown from the instructor and calculate percent error.

Sample Data Tables: *(Remember to prepare tables in your lab notebook before class.)*

Part I: Density of Wooden Block

	Trial 1	Trial 2
Length of wooden block (cm)		
Width of wooden block (cm)		
Height of wooden block (cm)		
Mass of wooden block (g)		

Part II: Density of a Metal
   Code # of Sample = _____

	Trial 1	Trial 2
Mass of metal pellets (g)		
Volume of water (mL)		
Volume of water with metal pellets (mL)		

Part III: Density of Unknown Liquid
   Code # of Sample = _____

Mass of empty grad cylinder (g)	
Mass of grad cylinder + liquid (g)	
Volume of liquid (mL)	

Pre-Lab Exercise:

1. What is the difference between precision and accuracy in the context of making measurements?

2. A student determined the mass of a graduated cylinder to be 31.4118 g. The correct value for the mass of the graduated cylinder is 31.4231 g. What is the % error in this measurement?

EXPERIMENT 2: DENSITY

3. Read and record the length of the block and the volume of the liquid in the graduated cylinder using the following diagrams.

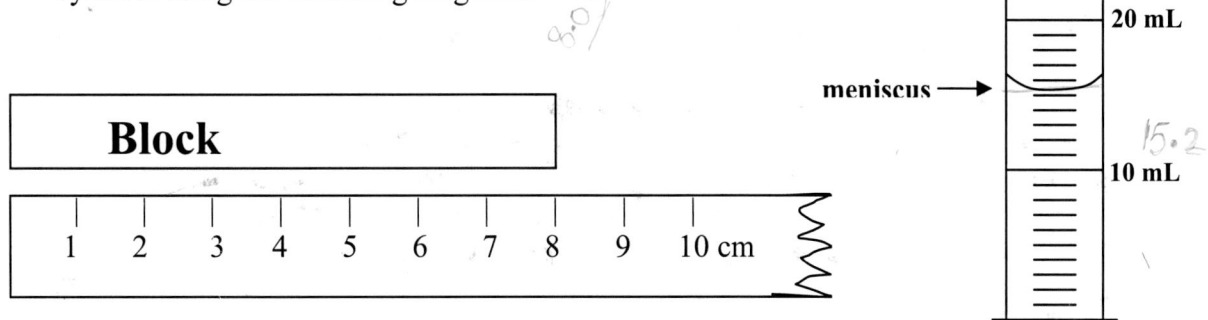

[handwritten: 8.0]

Block

1 2 3 4 5 6 7 8 9 10 cm

[graduated cylinder showing 20 mL and 10 mL markings, meniscus labeled]
[handwritten: 15.2 mL]

4. What is the mathematical formula for finding the volume of the following objects?
   a) Volume of a rectangular box;   b) Volume of a cylinder;   c) Volume of a sphere

   [handwritten: $V = whL$] *[handwritten: $V = \pi r^2 h$]* *[handwritten: $V = \frac{4}{3}\pi r^3$]*

5. The volume of water in a graduated cylinder was recorded as 31.9 mL. After a metal sample was immersed in it, the volume rose to 58.9 mL. What is the volume of the metal sample in the correct significant figures? If the metal weighs 72.8 g, what is its density?

6. In Part II, to minimize experimental error, what should you watch out for as you immerse the metal into the water? Describe at least two likely sources of error that you should avoid.

Post-Lab Questions: *[handwritten: Due next tuesday short lab]*

1. Does your experimental density of the metal sample agree with the true value? If no, give reasons for the discrepancy. Do not include operator errors.

2. Does your experimental density of the unknown liquid agree with the true value? If no, give reasons for the discrepancy. Do not include operator errors.

3. What is the density in g/mL of a ball that has a mass of 22.1 ounces and a diameter of 2.04 inches? Will it float if placed in a bucket of water? The answer is "no." Provide the calculations to prove this and explain in full sentences.

4. If there were air bubbles trapped under the metal pellets in Part II when you recorded its volume, would that give you an experimental density that is too high or too low? Explain. Would this be considered instrumental, method or operator error? Explain.

5. If you were to mix your unknown liquid in Part III with water, would you expect it to float or sink? Explain.

Experiment 3: CHROMATOGRAPHY

Purpose:
Part I: Separate and identify cations using paper chromatography
Part II: Separate and identify anions using thin layer chromatography

Performance Goals:
- Use chromatographic techniques to separate the components of a mixture
- Use chemical developing reactions to locate and identify components
- Calculate and use R_f values to identify components
- Make a representative sketch of a chromatogram

Introduction:

Chromatography is a physical method of separation that may be used both to separate and identify the components of a mixture. There are however, a variety of chromatographic techniques such as paper (PC), thin layer (TLC), column (CC), gas (GC), and high performance liquid chromatography (HPLC) that are in use. All of these techniques use a stationary phase and a mobile phase to effect separation. Each one uses different stationary and mobile phases with different types of interactions. This method of separation is based on the differences in the rates of migration of the components as they migrate through the *stationary phase* while being swept along by the *mobile phase*. Separation occurs because of differences in the affinity of each component for the stationary phase compared to the affinity for the mobile phase. These differences in affinity arise because of differences in composition and structure among the species being separated. Substances with a higher affinity for the stationary phase than for the mobile phase move more slowly as the mobile phase attempts to sweep them along while substances which have lower affinity for the stationary phase move more swiftly under the influence of the flow of the mobile phase. Similarly, components with high affinity for the mobile phase will move more swiftly than components which have a low affinity for the mobile phase. Simply put, strong interaction with the stationary phase tends to hold a component back while strong interaction with the mobile phase tends to move a component along.

For paper chromatography, the stationary phase is the paper and the mobile phase is a developing (or eluting) solvent. To separate a mixture a small spot of the mixture is placed near one end (*the origin*) of a piece of chromatography paper. That end of the paper is then placed in a developing solvent in the chromatography chamber. Capillary action sweeps the solvent over the stationary phase causing desorption and adsorption of the components. As the solvent creeps up over the paper the components that adsorb more strongly to the paper get left behind while the components that are less strongly adsorbed move ahead leading to a separation of the components. For samples that are colored, the component spots can be seen as the solvent creeps up the paper. For samples that are not colored, there are special techniques (discussed below) to cause them to become visible.

The paper is taken out of the solvent when the solvent reaches a designated distance near the top of the paper. If the components in a mixture all move at the same rate it means that there was no opportunity for any separation to occur and so another mobile phase should be tried.

EXPERIMENT 3: CHROMATOGRAPHY

In order to identify components of a mixture it is essential that they be completely separated from one another.

For thin layer chromatography, the stationary phase can be a thin layer of silica gel or other absorbent bound to an inert plastic, glass, or metal sheet. Similar to paper chromatography, the sample to be analyzed is spotted near the bottom (at the line of origin) of the TLC plate which is subsequently placed inside a chromatography chamber with a developing solvent. And like paper chromatography, the component spots can be seen if the sample is colored. This method of analysis is especially good for use in cases where only a small amount of material is present, for example, testing for the presence of drugs in urine and in blood.

If spots separated via paper chromatography or TLC are not visible, instrumental techniques or chemical developing reactions are used to locate the spots after the elution process. For this experiment, chemical developing reactions are used to make the invisible spots become visible. In the case of the paper chromatography, the paper will be sprayed with aqueous ammonia, followed by dimethyl glyoxime solution, and finally with sodium hydroxide solution. For the TLC, the TLC plate will be sprayed with ammoniacal silver nitrate solution followed by exposure to UV light. In each case the chemical sprays react selectively with the spots to form species that are colored and can be recognized by their colors. The chemicals in the spray solutions react differently with the different ions under investigation. For example, Ni^{2+} ions will react with dimethyl glyoxime to form a complex that has a bright pinkish red color.

Each component in a sample can be characterized by its *retention factor* (or R_f) value, which is the distance that a sample component has traveled from the line of origin, divided by the distance that the eluting solvent front has traveled from this same line (usually measured in cm). The geometric center of each spot is considered.

$$R_f = \frac{\text{distance traveled by component}}{\text{distance traveled by solvent}}$$

The R_f value for a sample component is independent of the size of the chromatography paper or TLC plate used in the separation. (If the solvent front moves further the sample component will also move further proportionally). R_f values will vary with different solvents and with paper or TLC plates of differing composition. If two or more components in a mixture have similar R_f values using a given solvent, they will not be separated (not be *resolved*) and thus cannot be identified.

Determination of the R_f Values:

Use a ruler to measure (in units of cm) the vertical distance each of the spots has moved from the origin line. Some of the spots might have become elongated and spread out as they rose from the origin line. In making these measurements, place a pencil dot in the **geometric center** of each spot and measure from the origin line to this dot, as shown in Figure 3.2. Also measure the distance the solvent moved from the origin line for each of the spots. Record these measurements in your lab notebook.

EXPERIMENT 3: CHROMATOGRAPHY

The unknown's may have several spots, each spot representing a different ion that is present in the unknown. Measure and record all spots that appear separately. Use these measurements to calculate the R_f values for each of the known ion spots and also the ion spots in the unknown's.

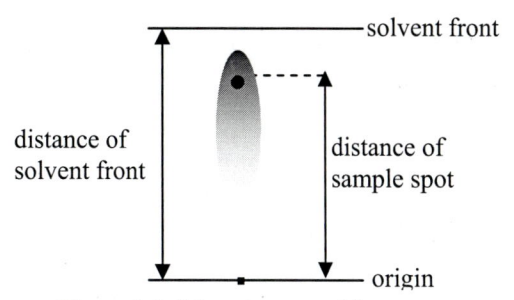

Figure 3.1: Measurement of Spot and Solvent Front

Part I: Separation and Identification of Cations by Paper Chromatography

Equipment/Materials:
Ruler (12-inch, graduated in cm), chromatography chamber (jar with lid or 300-mL beaker with Parafilm cover), developing solvent (acetone/6 M HCl in 9:1 ratio); chromatography paper (Whatman #1); capillary tubes; aqueous solutions of Co^{2+}, Cu^{2+}, Fe^{3+}, Mn^{2+}, Ni^{2+}; two unknown's, each containing one or more of the above cations; and the following solutions each in a spray bottle: aqueous ammonia solution, 1% dimethyl glyoxime (DMG) solution in 95% ethanol, 1 M NaOH solution

> **CAUTION: WEAR SAFETY GOGGLES AT ALL TIMES. SPRAYING MUST BE DONE IN THE HOOD. BE SURE TO WEAR GLOVES WHEN SPRAYING.**

Procedure: *In your lab notebook, set up a data tables similar to the one given at the end of this exercise.*

1. Pour 20 mL of the developing solvent into the chromatography chamber (jar or 300-mL beaker) and cover the chamber with the designated cover.
2. Obtain two unknown's from your instructor and record their unknown numbers in your notebook immediately. (You will receive a zero if you fail to record the unknown number.)
3. Use a pencil to make a faint line (called the origin) about 1.5 cm up along the long edge of the chromatography paper. No erasures are allowed as they would disrupt the fibers of the paper. (See Figure 3.2 below.)
4. Make a short line 5.0 cm from the line of origin. Later when the paper is placed inside the chromatography chamber, this line indicates how far the solvent should reach before the paper is to be taken out of the chamber.
5. Use separate capillary tubes to spot the known solutions (Co^{2+}, Cu^{2+}, Fe^{3+}, Mn^{2+}, Ni^{2+}) and your unknown's along the pencil line. The first and last spots should be about 2 cm from the short edges of the paper; and the spots should be about 1.5 cm apart. After a spot is dried, put another spot of the same solution right on top of it, to ensure that enough material is present.

46 EXPERIMENT 3: CHROMATOGRAPHY

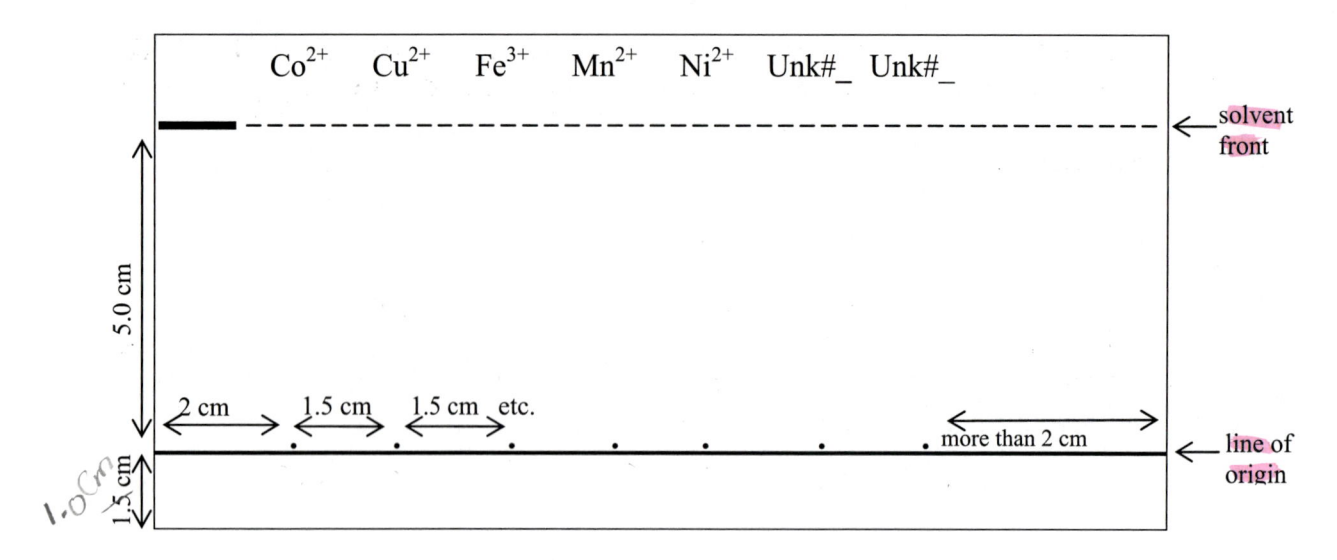

Figure 3.2: Preparation of the Chromatogram

6. Roll the short edges of the paper together (with the spots outside) to make a cylinder as shown here; and staple the edges. Make sure that the bottom corners are even and that the short edges do not overlap before fastening.
7. Place the cylinder with the spotted end down into the chamber and cover it without disturbing the solvent. Allow the solvent to move up, **undisturbed**, to the line marked as the solvent front at the top of the paper.

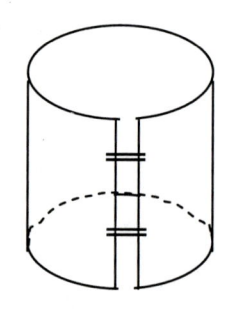

Do **NOT** move the chamber while the chromatogram is being developed.

8. Remove the paper from the chamber as soon as the solvent reaches the line marked as solvent front.
9. Use a heat gun to dry the paper. (Don't burn the paper—hold the heat gun about 6 in. from the paper.) The drying and subsequent spraying must be done in the fume hood.
10. Observe and record the colors of any spots which are visible.

CAUTION: REMEMBER TO WEAR GLOVES IN THE NEXT STEPS. SPRAYING MUST BE DONE IN THE HOOD.

11. Spray the paper with the aqueous ammonia solution, dry, observe, and record the effect on each spot.
12. Spray the paper with dimethyl glyoxime (DMG) solution, dry, observe, and record the effect on each spot.
13. Spray the paper with the NaOH solution, dry, observe, and record the effect on each spot.
14. Each unknown may contain up to 3 cations, each showing up as one spot after the chromatogram has been developed. Measure the distance from the origin to the solvent front and from the origin to the center of each spot. Use these values to calculate the R_f value for each spot.

15. It is possible that some of the spots will have R_f values that are very similar. Make use of both the colors and the R_f values of the separated cations to help you identify the cations present in your unknown's. State the identities in the lab notebook and in your conclusions.

16. Make a sketch of your chromatogram.

17. CLEANUP: Drain the used capillaries by spotting them on a practice piece of chromatography paper and place them in the broken glass container. Place the chromatography chamber in the place directed by your instructor. All reagent bottles must be returned to their original location.

Before arriving at the lab, prepare the tables shown below in your lab notebook:

Sample Data Table for Part I: Paper Chromatography

	Co^{2+}	Cu^{2+}	Fe^{3+}	Mn^{2+}	Ni^{2+}
Color of Sample					
Color of spots before developing					
Color of spots with NH$_4$OH					
Color of spots with DMG					
Color of spots with NaOH					
Distance traveled by component (cm)					
Distance traveled by solvent (cm)					
R$_f$ values					

For the unknown's, fill out the two tables as shown below. (Not all unknown's have 3 spots.)

Unknown # ___	Spot 1	Spot 2	Spot 3
Color of Sample			
Color of spots before developing			
Color of spots with aqueous ammonia			
Color of spots with DMG			
Color of spots with NaOH			
Distance traveled by component (cm)			
Distance traveled by solvent (cm)			
R$_f$ Values			

Unknown # ___	Spot 1	Spot 2	Spot 3
Color of Sample			
Color of spots before developing			
Color of spots with aqueous ammonia			
Color of spots with DMG			
Color of spots with NaOH			
Distance traveled (cm)			
R$_f$ Values			

Conclusion: Unknown # ___ contains _____.
Unknown # ___ contains _____.

Part II: Separation and Identification of Anions by Thin Layer Chromatography

Equipment/Materials:

Chromatography chamber: jar with lid or 250-mL beaker with aluminum foil (10 cm x 10 cm as cover), silica gel TLC plate (7 cm x 5 cm), capillary tubes, developing solvent (s-butanol/methyl ethyl ketone/aqueous ammonia in 20:20:10 ratio), ammoniacal silver nitrate, known solutions (Cl$^-$, Br$^-$, I$^-$), mixture of all three, one unknown sample containing one or more of the above anions, UV chamber

> **CAUTION: WEAR SAFETY GOGGLES AT ALL TIMES. SPRAYING MUST BE DONE IN THE HOOD. BE SURE TO WEAR GLOVES WHEN SPRAYING. AVOID INHALATION OF THE FUMES FROM THE SOLVENT.**
> *Silver nitrate will darken your skin and the spots will remain on your skin until your skin wears off (in a couple of days).*

Procedure: *Work individually. In your lab notebook, setup a data table similar to the one shown below. Write neatly and legibly.*

1. Measure the volume specified by your instructor, of the developing solvent in the chromatography chamber and cover it with its lid. (If a 250-mL beaker is used as the chamber, measure 7 mL and cover with the aluminum foil provided.) While the chamber is being saturated by the vapors of the solvent, proceed to the next step.
2. Obtain an unknown from your instructor and record its unknown number in your lab notebook immediately. (You will receive a zero if you fail to record the unknown number.)
3. Obtain a TLC plate from the side shelf. Remember to hold it only by the side or top edge and place it on a clean and dry surface.
4. Use a **pencil** to draw a very **faint** line about 1.20 cm from the bottom of the TLC plate. **Do not press hard, it may cut into the absorbent.** (See figure below.)
5. Faintly mark five spots on the origin line with your pencil. The first spot should be 0.50 cm from the side and the rest of the spots should be 0.80 cm apart.
6. Mark a short line 4.50 cm above the line of origin. This helps you determine when to lift the TLC plate out of the chamber.

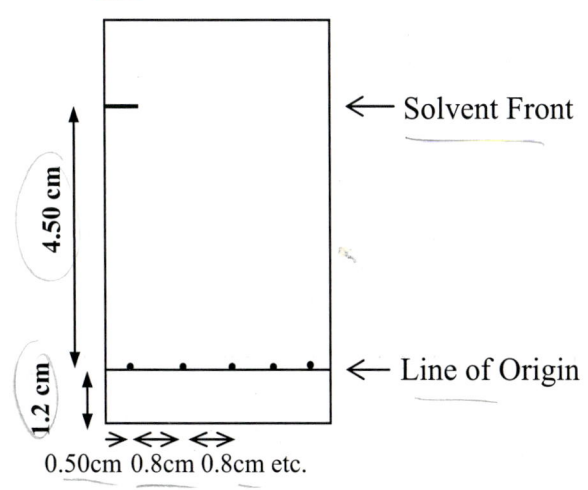

50   EXPERIMENT 3: CHROMATOGRAPHY

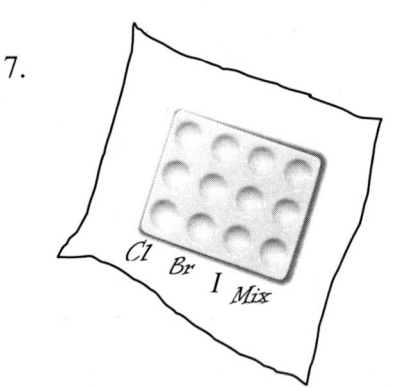

7. Obtain a ceramic spot plate (see figure on the left). Clean and dry it if necessary. Place it on a piece of paper and label four of the wells as shown. Add one or two drops of each known halide and the mixture (contains all three halides) to separate wells. The unknown does not have to be transferred to the well plate and can be accessed directly from the vial.

8. Using separate capillary tubes, spot each of the sample on the adsorbent side of the TLC plate. Spot them in this order: Cl⁻, Br⁻, I⁻, the mixture, and your unknown. What you need is a very small but concentrated spot. The proper technique is to touch the capillary tube to the sample on the spot plate. Capillary action will draw a small sample up the tube. Holding the tube vertical, touch the plate on the marked spot **very briefly**. If you hold it on the TLC plate too long, the spot would become too big. Repeat this process again, spotting the same sample exactly on the same spot. This technique gives you a very small but concentrated spot.

   Repeat the process with another halide, following the order stated above. Be careful not to cross contaminate by using the same capillary for another samples.

   For the mixture and for the unknown, spot each sample three times instead of twice. These solutions are less concentrated than the known's and need to be spotted an extra time.

9. After the spots have dried, carefully place the TLC plate into the chromatography chamber and quickly cover with its lid without disturbing the solvent. **ONCE THE TLC PLATE IS IN THE CHROMATOGRAPHY CHAMBER, IT IS CRUCIAL THAT YOU DO NOT SLOSH THE SOLVENT AROUND!** Disturbing the solvent will result in the spots traveling at an angle instead of vertically. The spots should be at the bottom and the backing (not the adsorbent) should rest against the wall of the chromatography chamber. Check to make sure that the line of origin is above the level of the solvent in the beaker.

10. Allow the solvent to move up to the top of the plate where you had placed the 4.5 cm mark. While you are waiting, with the help of a ruler, begin making a sketch (**to scale**) of your chromatogram in your notebook.

11. Remove the TLC plate and quickly trace the solvent front with your pencil (i.e. the position that the solvent reached on the slide). Use the heat gun to dry the plate. Use the hot but MEDIUM setting and hold the heat gun at least 6 inches from the plate. If the hot air is too close, you could blow part of the adsorbent off the plate!

12. Record the color of the spots before spraying.

13. Put on your gloves if you are not already wearing them. Go to the designated hood and hold the TLC plate securely with crucible tongs. Spray the TLC plate with ammoniacal silver nitrate and dry again. **Do not get this spray on your hands**.

14. Record the color of the spots after drying.

15. Place the plate in a UV chamber and using the short wave rays, watch for spots to develop. Mark the spots with a pencil. Make note of any differences in the colors of the spots. This may aid you in the identification of your unknown.

EXPERIMENT 3: CHROMATOGRAPHY

16. Calculate the R_f values of each spot in the known's, mixture and unknown and identify the components of your unknown accordingly.
17. Complete your sketch of the chromatogram by adding the spots and labeling them.
18. CLEANUP: Discard the capillary tubes by placing them in the broken glass container. Your instructor will specify what to do with the developing solvent and unknown vial.

Sample Data Table for Part II: Thin Layer Chromatography

	Cl⁻	Br⁻	I⁻	Mixture of all three
Color of samples				
Color of spots before spraying				
Color of spots after spraying & drying				
Distance traveled by spot (cm)				Spot 1
		N/A		Spot 2
	N/A		N/A	Spot 3
Distance traveled by solvent (cm)				
R_f values				Spot 1
	N/A	N/A	N/A	Spot 2
	N/A	N/A	N/A	Spot 3

Unknown Code # ___
Color of unknown sample:

	Spot 1	Spot 2	Spot 3
Color of spots before spraying			
Color of spots after spraying & drying			
Distance traveled by spot (cm)			
Distance traveled by solvent (cm)			
R_f values			

CONCLUSION: Unknown #___ contains _____.

EXPERIMENT 3: CHROMATOGRAPHY

Pre-Lab Exercise:
Your instructor will indicate whether you are to submit answers to the following questions at the beginning of the pre-lab discussion or you are to take a quiz with similar questions.

1. Read the introduction to learn how to calculate R_f values such as the ones for B1 and B2 in the section of a chromatogram shown below. (Show work and watch your sig. fig.)

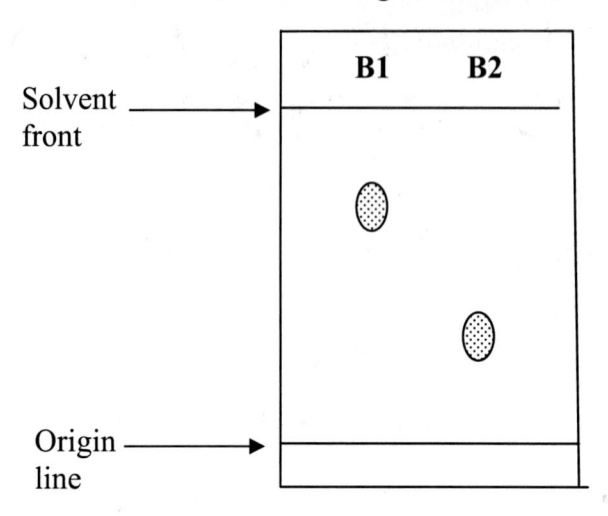

2. Below are three chromatograms that were developed in different solvents (X, Y, and Z). Which of these three solvents would be best suited for identifying the substance R40 in an unknown sample that might also contain one or more of the substances shown on these chromatograms? Explain your answer.

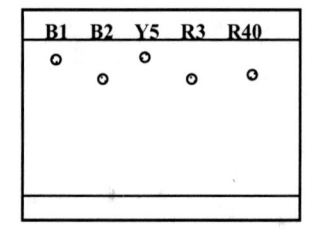

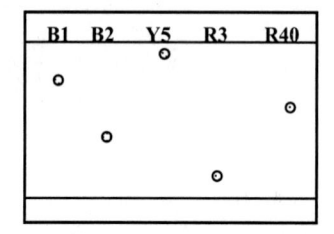

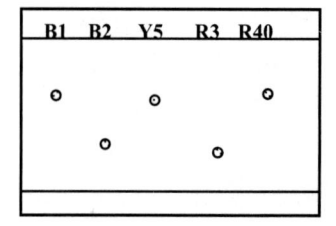

   eluted with Solvent X   **eluted with Solvent Y** ✓   **eluted with Solvent Z**

3. Based on the three chromatograms above, if you used Y as the eluting solvent, and unknown #1 contains both B1 and B2, and unknown #2 contains both Y5 and R40, what would your chromatogram look like after it has been developed? Answer by sketching a chromatogram similar to the one below and include the spots you would expect to see.

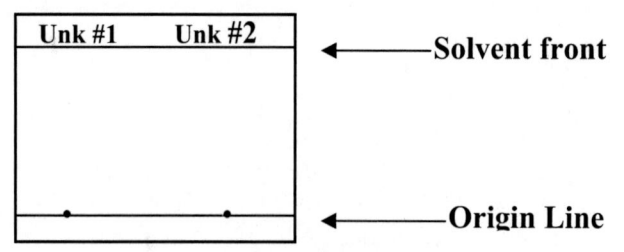

EXPERIMENT 3: CHROMATOGRAPHY

Post-Lab Questions: (Answers must be in full sentences and typed.)

1. Suppose you were not paying attention to your chromatogram and the solvent rose all the way to the top of the paper or TLC plate and remained that way for a period of time before you removed it from the chromatography chamber. Suppose none of the spots have reached the top yet and you assumed the solvent front to be at the top paper or TLC plate. How would this mistake affect R_f values of the ions? Would the R_f values be too high, too low or unaffected? Explain. *(Note: When the solvent reaches the top edge of the paper or TLC plate, vaporization will allow the solvent to continue rising up the paper.)*

2. If measurements made were in inches instead of centimeters, how would the R_f values be affected? Explain.

3. Would any of the ions be hard to differentiate if color alone were used? Explain.

4. Would any of the ions be hard to differentiate if R_f alone were used? Explain.

5. a) In the paper chromatography, based on your R_f values, which ion has the highest affinity for the eluting solvent? Explain.

   b) In the thin layer chromatography, based on your R_f values, which ion has the highest affinity for the eluting solvent? Explain.

54   EXPERIMENT 3: CHROMATOGRAPHY

Experiment 4:
COMPOSITION OF A HYDRATE

Purpose: Determine the empirical formula of an unknown hydrate and the percentage by mass of water in the hydrate

Performance Goals:
- Gain skills in the operation of the Bunsen or Fisher burner
- Gain skills in handling hot apparatus in the laboratory
- Gain skills in drying to constant mass
- Perform calculations to determine the empirical formula of an unknown hydrate
- Perform calculations to determine the mass percent of water in an unknown hydrate

Introduction:

Certain compounds form crystals with a definite proportion of water molecules incorporated in the crystal structure. These compounds are called *hydrates*. Copper(II) sulfate pentahydrate is an example of such a hydrate. Its formula is $CuSO_4 \cdot 5H_2O$. The five in front of the formula for water tells us there are 5 water molecules per formula unit of $CuSO_4$ (or 5 moles of water per mole of $CuSO_4$). The water in the formula is referred to as the *water of hydration*, and the dot indicates that the water is chemically bonded to the $CuSO_4$ salt. (It does not represent the multiplication sign.)

Note that this water of hydration is very different from the water contained within a sample that is merely wet. A wet sample can have a variable amount of water regardless of the formula of the compound. In contrast, a hydrate contains a specific number of moles of water of hydration per mole of the salt. Some compounds form more than one hydrate, each containing a different number of moles of water of hydration.

When heated above 100°C, hydrates lose their water molecules fairly easily since the hydrate bonds are much weaker than the ionic bonds between the salt ions. When the water is driven off, the resulting product is the anhydrous salt which is called an *anhydride*.

For example: $CuSO_4 \cdot 5H_2O(s) \xrightarrow{\Delta} CuSO_4(s) + 5H_2O(g)$
                       hydrate   anhydride
                                               or anhydrous salt

The hydrated salt loses mass as the anhydride is formed. When no further mass loss is observed the process is complete. A change in appearance from crystalline to powdery form (sometimes accompanied by a change in color) may be observed as the process proceeds.

The primary objective of this experiment is to use experimental data to determine the *empirical formula* of an unknown hydrate. The formula derived from experimental data is called an *empirical formula*. The term, *empirical*, means that it is based on observed data rather than on theory. It can give us only the lowest ratio of the components in a compound unless other information is available.

EXPERIMENT 4: COMPOSITION OF A HYDRATE

In this experiment a weighed sample of an unknown hydrate will be heated and the mass of water lost on heating will be determined. The formula of the anhydride is provided and based on the moles of the hydrate present before heating and the moles of the water lost upon heating, the empirical formula of the hydrate can be determined. In addition, the mass percent of water lost by the hydrate can be calculated.

Equipment/Materials:
Crucible (without lid), clay triangle or wire triangle, iron ring, ring stand, lab burner (Bunsen or Fisher), crucible tongs, electronic balance, heat pad, approx. 3 grams of an unknown hydrate (in plastic bags or bottles)

Procedure:
The experiment will be performed individually, and two trials are required.
All masses must be recorded to 0.001 g.

1. Obtain a sample of an unknown hydrate. The formula of the unknown anhydride is provided. Record it in the lab notebook.

2. Obtain a crucible and support the crucible on a clay or wire triangle. If there is residue inside the crucible left over from a previous experiment, dump it out into the special waste container in the hood and wipe the inside and outside with dry paper towel. Do not wash it. It will prevent introduction of any moisture to the crucible.

3. Adjust the burner to give a flame with a blue inner cone by opening up the chimney to allow more air in. A luminous, yellow flame is not hot enough, and in addition, will produce undesirable soot that will collect on the bottom of the crucible. Position the burner so that the hottest part of the flame, found at the tip of the inner blue cone, is at the bottom of the crucible. When set up properly, the bottom of the crucible should be glowing red hot within a couple of minutes. Obtain help from the instructor if it does not glow red.

4. Continue heating for 5 minutes. This will remove all moisture that may have been present in the crucible.

5. Turn off the burner and allow the crucible to cool for about 10 minutes until it is at room temperature. A warm object will appear to weigh less than its true weight because its heat will create a convection current that affects the operation of the balance.

6. Make use of crucible tongs and the wire gauze to transfer the crucible safely to the balance. The proper way to hold the crucible is to cradle it with the crucible tongs as shown in Figure 4.1. This technique avoids introducing contaminants from the tongs to the inside of the crucible, and accidentally removing some of the residue from the crucible as shown in Figure 4.2.

EXPERIMENT 4: COMPOSITION OF A HYDRATE   57

Figure 4.1: Correct Way to Hold a Crucible with Crucible Tongs

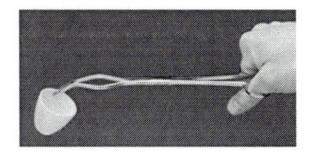

Figure 4.2: Crucible Tongs Can Introduce Contaminants

7. Record the weight of the cooled crucible as "Mass of crucible" under Trial 1.

8. ***There is enough unknown sample for two trials.*** Transfer about 1.5 grams of your hydrate (from the plastic bag or bottle) into the crucible. Record the weight of the crucible as "Mass of crucible and hydrate" under Trial 1.

9. Place the crucible (without the lid) with the hydrate on the clay or wire triangle. Heat gently for about 3 minutes by holding the base of the burner in hand and playing the flame back and forth on the bottom of the crucible. Vigorous heating when the hydrate still contains most of its water may cause it to splatter and introduce a significant error into the experiment.

10. Next heat strongly for another 5 minutes. The crucible bottom should be glowing red hot during this heating period. Allow the crucible and its contents to cool for about 10 minutes. Record its weight as "Mass of crucible and anhydride (after 1st heating)".

11. Repeat the heating at maximum temperature for an additional 6 minutes; cool the crucible and record its weight as "Mass of crucible and anhydride (after 2nd heating)". In this second and subsequent heating, it is not necessary to do a preliminary gentle heating as the sample no longer contains an excessive amount of water.

12. If the decrease in mass between these two weighings (weight after 1st heating and after 2nd heating) is more than 5 mg (0.005 g), repeat the heating (as in step 11) once more and record the weight for the 3rd heating. The idea is to reach a constant, final mass to ensure all of the water has been removed.

13. Dump the contents of the crucible into the special waste container in the hood and wipe the inside and outside with dry paper towel. Again, do not wash it. Weigh the empty crucible and record this in your lab notebook under Trial 2.

14. Add the remaining hydrate (about 1.5 g) to the crucible. Record the weight as "Mass of crucible and hydrate" under Trial 2.

EXPERIMENT 4: COMPOSITION OF A HYDRATE

15. Repeat the heating, cooling and weighing sequences described for Trial 1 (Steps 8 through 12). Remember to heat the new hydrate <u>gently</u> the <u>first</u> time before heating it vigorously, and to repeat the sequence until a constant final mass is obtained.

16. CLEANUP: Discard all of the residue in the designated container in the hood. Wipe the inside of the crucible with a dry paper towel and return it to the side shelf.

Calculations:

1. Calculate the mass of the hydrate. It is the difference between the mass of the crucible containing the hydrate and the mass of the empty crucible.

2. Calculate the mass of the anhydride. It is the difference between the <u>final</u> mass of the crucible <u>after the last heating</u> and the mass of the empty crucible.

3. Calculate the mass of water lost on heating. It is the difference between the mass of the hydrate and the mass of the anhydride.

4. Calculate the mass percent of water in your hydrate for each trial & the average mass %.

$$\text{Mass \% water in hydrate} = \frac{\text{Mass of water}}{\text{Mass of hydrate}} \times 100$$

5. Calculate the molar mass for your anhydride from the formula provided.

6. Calculate the number of moles of anhydride from its molar mass and your mass of anhydride.

7. Calculate the number of moles of water of hydration from the molar mass of water and the mass of water lost on heating.

8. Determine the number of moles of water of hydration per mole of anhydride. This is done by calculating the ratio of moles of water to moles of anhydride. (Watch sig. fig.!)

$$\text{Ratio of moles of water per mole of anhydride} = \frac{\text{\# mol water}}{\text{\# mol anhydride}}$$

Round the ratio to the nearest whole number. This number gives us the number of water molecules per formula unit of the anhydride. It is the x in the empirical formula of the hydrate: anhydride·xH_2O.

9. Using this ratio, write the empirical formula of your hydrate.

10. Calculate the theoretical percent of water based on the empirical formula obtained in step 9 and then calculate the percent error in the mass percent of water found in step 4. (Review equations in Experiment #2 on the calculations of error and % error.)

EXPERIMENT 4: COMPOSITION OF A HYDRATE 59

Sample Calculations: *The instructor will go over these calculations at the beginning of the lab. Students are encouraged to complete these calculations before class to be better prepared for the experiment.*

*This is an example of the calculations you will be doing for this experiment. The hydrate used in this example is **NOT** the same as the given unknown. Be familiar with the terms, hydrate, anhydride, water of hydration.* $CuSO_4 \cdot 5H_2O$ *is a hydrate, with five molecules of water of hydration. Its corresponding anhydride is* $CuSO_4$. *In this calculation, the hydrate is* $FeCl_3 \cdot xH_2O$ *where x is the unknown number of moles of water of hydration in the formula that we are trying to determine.*

Data:
Given: Formula of the anhydride = $FeCl_3$
1. Mass of crucible = 18.754 g
2. Mass of crucible + hydrate = 20.334 g
3. Mass of crucible + anhydride (after 1st heating) = 19.722 g   ← Compare these 2 masses. What do they tell you?
4. Mass of crucible + anhydride (after 2nd heating) = 19.705 g
5. Mass of crucible + anhydride (after 3rd heating) = 19.704 g
   (only if necessary)

Mass of hydrate = 20.334 g − 18.754 g = 1.580 g $FeCl_3$

Mass of anhydride = 19.704 g − 18.754 = 0.950 g anhydride

Mass of water lost on heating = mass of hydrate − mass of anhydride
= 1.580 g − 0.950 g = 0.630 g H_2O

% Percent water in hydrate by mass = $\dfrac{\text{mass of } H_2O}{\text{mass of hydrate}} \times 100$ = $\dfrac{0.630 \text{ g } H_2O}{1.580 \text{ g } FeCl_3} \times 100$ = 39.9 % H_2O

Moles of water of hydration = 0.630 g H_2O × $\dfrac{1 \text{ mol}}{18.02 \text{ g}}$ = 0.0350 mol H_2O

Moles of anhydride = 0.950 g × $\dfrac{1 \text{ mol}}{162.2 \text{ g}}$ = 5.86 × 10⁻³ mol of anhydride

Ratio of moles of water to moles of anhydride = $\dfrac{\text{mole of water}}{\text{mole of anhydride}}$ = $\dfrac{0.0350 \text{ mol } H_2O}{0.00586 \text{ mol of anhydride}}$

= 5.97 ≈ 6 H_2O

Empirical Formula of the hydrate = $FeCl_3 \cdot 6H_2O$

Theoretical % water based on empirical formula = $\dfrac{\text{mass of } H_2O}{\text{mass of anhydride}}$ = $\dfrac{6(18.02)}{270.3 \text{ g anhydride}} \times 100$
= 40.00 % H_2O

Error in % water = $\dfrac{\text{expt} - \text{theoretical}}{\text{theo}} \times 100$ = $\dfrac{39.9 - 40.00}{40.00} \times 100$ = −0.25 %

Percent Error in % water =

EXPERIMENT 4: COMPOSITION OF A HYDRATE

Prepare this Data Table in your lab notebook before arriving at the lab:

Sample Data Table

	Trial #1	Trial #2
Formula of your salt (anhydride)		
Mass of crucible		
Mass of crucible + hydrate		
Mass of crucible + anhydride (after 1st heating)		
Mass of crucible + anhydride (after 2nd heating)		
Mass of crucible + anhydride (if necessary, after 3rd heating)		

Pre-Lab Exercise:

1. Write a <u>balanced</u> equation for the decomposition by heating of iron(II) sulfate heptahydrate. *(Hint: Use your chemistry text to help you determine the formula of iron(II) sulfate heptahydrate.)*
2. Calculate the molar mass of iron(II) sulfate heptahydrate in 4 sig. fig.
3. Calculate the theoretical % of H_2O in iron(II) sulfate heptahydrate in 4 sig. fig.
4. In the reaction described in Question #1,
   a) What is the formula of the hydrate? b) What is the formula of the anhydride?
   c) How many molecules of water of hydration are in each formula unit of the hydrate?
5. A sample of hydrate is being heated in a crucible to determine its water content as described in the experiment. What is the criterion for deciding whether a third heating is necessary? Be specific and explain your answer.
6. What is the proper way of using crucible tongs, and why is it best done that way?

Post-Lab Questions: (Answers must be in full sentences and typed.)

1. In this experiment you are to remove the water of hydration by heating. Why do you have to heat it at least twice instead of heating the hydrate <u>just once for a prolonged period of time</u>?

2. If you neglected to heat the hydrate <u>gently</u> the first time and splattering occurs, how would it affect your percent of water of hydration? Be specific. Predict whether it would be too high or too low and explain your answer.

3. A student, John, reported the following for his unknown hydrate:

	Trial 1	Trial 2
% water	36.5 %	36.2 %
Water of Hydration (before rounding)	4.02	4.03

   We know there is something wrong with his reported values even without knowing the formula of his anhydride. What type of error would cause this? Explain fully.

4. A student, Mary, used 1.2 g of hydrate in the first trial and 1.6 g in the second trial. Explain fully why her mass percent of water will be the same for both trials even though she used more hydrate in the second trial. *(Do not simply state that the mass percent is an intensive property.)*

Experiment 5: SYNTHESIS OF TRIS(ETHYLENEDIAMINE)NICKEL(II) CHLORIDE

Purpose: Synthesize a nickel(II) complex and apply reaction stoichiometry to determine the percent yield

Performance Goals:
- Prepare a complex compound
- Identify the limiting reactant in a reaction
- Determine theoretical yield
- Determine percent yield

Introduction:

Complex compounds also known as coordination compounds are formed when molecules or ions bond to metal ions to form more complex structures. The molecules or ions that become attached to a metal ion are called ligands. Ligands must contain at least one unshared electron pair that can be donated to the metal ion to form a metal-ligand bond which is called a coordinate covalent bond.

The synthesis reaction studied in this experiment is represented by the balanced equation:

$$NiCl_2 \cdot 6H_2O(s) + 3H_2NCH_2CH_2NH_2(aq) \longrightarrow [Ni(H_2NCH_2CH_2NH_2)_3]Cl_2(s) + 6H_2O(l) \quad \text{Equation 1}$$

"hydrate"   ethylenediamine or "en" $Ni(en)_3Cl_2$ or "Tris"

The equation shows that three moles of ethylenediamine, abbreviated **en**, are necessary to react with one mole of nickel(II) chloride hexahydrate, abbreviated **hydrate**, to form one mole of the complex compound, tris(ethylenediamine)nickel(II) chloride, abbreviated as **Ni(en)₃Cl₂** or **Tris**. The ethylenediamine (en) molecule acts as the ligand in this reaction and because it bonds to the nickel ion in two different positions, it is called a *chelating ligand*. The word "chelate" has Greek and Latin origins referring to a claw-like or pincer action. In this reaction each nitrogen atom (using its lone pair of electrons) in the en molecule bonds to the nickel ion; and there are three en molecules per nickel ion, forming the $Ni(en)_3^{2+}$ complex ion. The chloride ions in the solution, $Cl^-(aq)$, form ionic bonds with the complex ion giving a purple, crystalline solid which precipitates from the solution. The structures of the complex and the ligand are shown below.

Ethylenediamine (en) is represented here as

$H_2N:\quad :NH_2$

Ni(en)₃Cl₂ or Tris

When synthesizing chemical compounds from two (or more) reactants, the usual practice is

63

EXPERIMENT 5: SYNTHESIS OF Ni(en)₃Cl₂

to add an excess amount of one (or more) reactants so that one reactant is completely used up. Often one of the reactants is available in small quantity due to the amount available or cost. The other reactant(s) is/are intentionally placed in the reaction flask in excess quantity to force reaction to completion.

The reactant which is completely used up limits the maximum amount of each product which can be produced. This reactant is called the limiting reactant and the amount of each product that can form from it is referred to as the *theoretical yield*. The theoretical yield is the maximum quantity of product that can be obtained (in terms of moles or grams) based on the amount of the limiting reactant used by making use of reaction stoichiometry. Usually the quantity of product actually isolated in the reaction (the *actual yield*) is less than the theoretical yield. Generally, the quantity that is reported in a chemical synthesis reaction report is the percent yield. The percentage yield of a product can be determined as follows:

$$\text{Percent yield} = \frac{\text{Actual Amount of Product Obtained}}{\text{Theoretical Amount of Product Expected}} \times 100 \qquad \text{Equation 2}$$

In this experiment, both reactants and the complex product are water soluble. We begin with the reaction taking place in water. Acetone is then added to precipitate out the Ni(en)₃Cl₂ product which is much less soluble in this acetone/water mixture. And, because the reactants are soluble, any excess reactant remains in solution. It is important that a minimal amount of water is used in the synthesis to maximize the amount of Ni(en)₃Cl₂ that precipitates out.

Equipment/Materials:

100-mL or 150-mL beaker, spoonula or Scoopula, electronic balance, glass rod/rubber policeman, watch glass, dropper, 10-mL graduated cylinder, filter flask, Büchner funnel, gasket, filter paper, vacuum hose (heavy-walled), ice-bath, ring stand, clamp/clamp fastener, infrared lamp, plastic envelope or vial, nickel(II) chloride hexahydrate, de-ionized water, aqueous 25.0% ethylenediamine solution in a dropper bottle, and acetone

Procedure:
Prepare a data table in your lab notebook similar to what is shown below before arriving at the lab.

1. *Tare* an empty, dry 100- or 150-mL beaker to zero. (This means zero the balance as usual, then place the beaker on the pan and press the TARE button to re-set it to zero.)
2. Remove the beaker from the balance and with your spoonula or Scoopula add a small amount of nickel(II) chloride hexahydrate (NiCl₂·6H₂O) to the beaker, place it back on the balance pan and check the mass. Repeat this process until the balance reads between 0.500 and 0.650 g. (Since the empty beaker was tared to zero, the balance reading represents the mass of the hydrate alone.) Record the exact mass and note the color and crystal form of the compound in your lab notebook.
3. With the use of a dropper slowly add deionized water drop by drop, mixing after each drop, until the solid is completely dissolved. Use your glass rod (not the end with the rubber policeman) to grind up large pieces of the hydrate. Remember not to use excess

EXPERIMENT 5: SYNTHESIS OF Ni(en)₃Cl₂   65

water to dissolve the solid because the product of the reaction is soluble in water. Note the color of the solution in your lab notebook.

4. Use a 10-mL graduated cylinder to measure out 2.3–2.5 mL of 25.0% ethylenediamine solution. Record the volume to the nearest 0.1 mL, and the appearance of the en solution. Slowly add the entire solution to the beaker while constantly mixing. This solution has a density of 0.950 g/mL and is 25.0% ethylenediamine by mass. Record this information in your lab notebook. Feel the outside surface of the beaker. Record what else tells you a reaction is occurring.

> **CAUTION: The acetone used in the next step has a very low boiling temperature. It is easily absorbed through the skin. Both the liquid and vapors are extremely flammable. Extinguish all flames before using acetone and avoid contact with your skin. It will also remove nail polish.**

5. Add 15.0 mL of acetone in 5-mL increments, mixing well after the addition of each aliquot.
6. Use a glass stirring rod to stir the mixture until precipitation begins. This may require a few minutes of vigorous scratching on the inner wall of the beaker. Record your observations.
7. After precipitation begins, allow the beaker to sit on the desktop for a few minutes to cool to near room temperature.
8. Cool the beaker further in a slush bath made of 2 parts ice to 1 part water for at least 15 minutes to maximize precipitation of the product. The product is ionic and therefore soluble in water, but it is not soluble in the less polar acetone, especially when cold. Do not let the beaker tip over! There is no time to start over.
9. Set up a suction filtration apparatus as shown in the diagram below. Place a piece of filter paper inside the Büchner funnel. Be sure the filter flask and the Büchner funnel are clean.

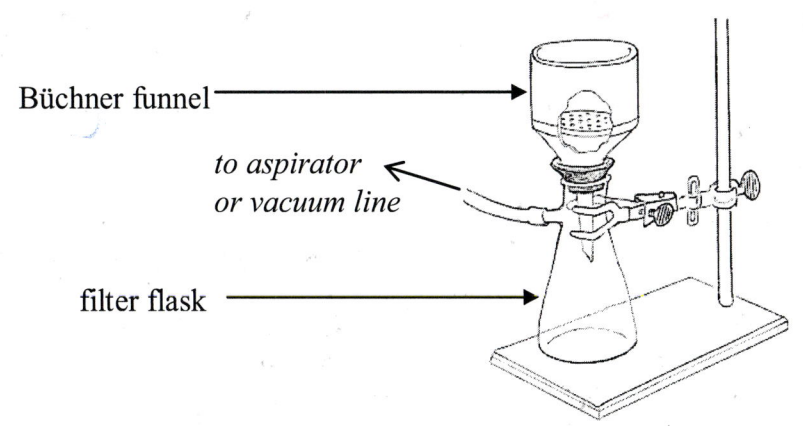

Büchner funnel

to aspirator or vacuum line

filter flask

10. Start the suction and wet the filter paper with a quick squirt of water. (This will help the filter paper form a seal thus preventing solids from seeping underneath the edges of the paper.) Then transfer the solid/liquid mixture to the funnel. If some of the precipitate gets around or through the filter paper, consult with your instructor. A small amount of precipitate in the filter flask is permissible. If a significant amount is present, turn off

the suction, pour the filtrate back into the beaker, and re-filter as before. To retrieve any precipitate that remains in the beaker, add a few mL of cold acetone to the beaker, scrape the particles loose with a rubber policeman on the end of a stirring rod, and transfer them into the Büchner funnel.

11. Break the suction by removing the hose from the sidearm of the filter flask then turn off the suction. Break up the solid with a spoonula or Scoopula, being careful not to puncture or tear the filter paper. After breaking up the solid, re-attach the hose to the sidearm of the filter flask and turn on the suction again. While the suction is on, wash the precipitate by slowly pouring a 5-mL portion of cold acetone over the surface of the precipitate. Keep the suction on for at least 5 minutes (the longer the better) to dry the precipitate.

12. Break the suction again and then turn off the suction. Remove the Büchner funnel and carefully scrape the crystals from the filter paper onto a clean and dry watch glass. Use your spoonula or Scoopula to spread the crystals out and break up any lumps that may be present. Place the watch glass with the crystals under an infrared lamp for at least 15 minutes to complete the drying process, the longer the better. ***Do not adjust the height of the heat lamp.*** Pour the wash liquids from the filter flask into the waste container in the hood.

13. Weigh a plastic envelope (or vial) labeled with your name, date, lab section and Ni(en)$_3$Cl$_2$ for the name of the compound. Transfer the product into the envelope (or vial) and reweigh. (Remember to use the same balance.) Record the appearance of your product in your lab notebook.

14. Determine the mass of product by difference. *What does that mean you have to do?* Put the mass of your product on the label also. Your instructor will specify whether you are to turn in the sample at the end of the period or with a lab report due the following week. Do not take the product home.

Calculations:

The goal is to determine the theoretical yield and then the percent yield of the product. For simplicity sake, the reactants and product will be referred to by their nicknames:

HYD = NiCl$_2$·6H$_2$O   EN = H$_2$NCH$_2$CH$_2$NH$_2$ TRIS = [Ni(H$_2$NCH$_2$CH$_2$NH$_2$)$_3$]Cl$_2$

1. In order to calculate the theoretical yield, you must first determine whether HYD or EN is the limiting reactant. Begin with determining the number of moles of each reactant used.
    a. For the reactant, HYD, the number of moles is calculated from its mass and its molar mass.
    b. For the other reactant, EN, it is less simple. Since pure EN is a thick, viscous, strongly alkaline liquid, it is commonly used in solution rather than as a pure liquid. What you recorded was the volume of the EN solution. The mass (m) of the EN solution is determined from the volume (V) and density (d) of the solution (Equation 3). The mass of the pure EN is calculated from the mass of the solution and the mass percent of EN in the solution (Equation 4). Once you have the mass of pure EN, you can convert it to number of moles of EN. The solution used in this experiment is 25.0% by mass EN with a density of 0.950 g/mL.

EXPERIMENT 5: SYNTHESIS OF Ni(en)$_3$Cl$_2$

$$m(EN\ soln) = d(EN\ soln) \times V(EN\ soln) \qquad \text{Equation 3}$$

$$(\text{Mass pure EN})\ g\ EN = m(EN\ soln)\ g\ \cancel{EN\ soln} \left(\frac{25.0\ g\ EN}{100\ g\ \cancel{EN\ soln}} \right) \qquad \text{Equation 4}$$

Now that you have the number of moles of each reactant, you can then determine the number of moles of TRIS product expected based on each reactant using the coefficients of the balanced equation. The reactant that yields the least amount of product is the limiting reactant (because it will run out first and therefore limit the amount of product formed). The number of moles of product based on the limiting reactant is the **theoretical yield in moles**. Convert this to grams using the molar mass of the TRIS product and you have the **theoretical yield in grams**.

Finally, determine the percent yield of TRIS, using Equation 2.

Before arriving at the lab, prepare a data table shown below in your lab notebook:

Sample Data Table

Mass of NiCl$_2$·6H$_2$O = _____ g
(Obtained using the TARE feature)
Appearance of solid NiCl$_2$·6H$_2$O
Appearance of NiCl$_2$·6H$_2$O solution
Volume of ethylenediamine solution (en soln) = _____ mL
Concentration of en solution (From the label on the bottle) = _____ mass %
Density of EN solution (From Procedure Step 4) = _____ g/mL
Appearance of en solution **before** addition to the hydrate:
Appearance of reaction mixture **after** addition of en solution to the hydrate:
Appearance of reaction mixture **after** addition of acetone:
Color of product, Ni(en)$_3$Cl$_2$, after drying process:
Mass of empty plastic bag (or vial) + product = _____ g
Mass of empty plastic bag (or vial) = _____ g
Mass of product (actual yield) = _____ g

Pre-Lab Exercise:

1. To determine the molar mass of the reactants would you calculate it for H$_2$NCH$_2$CH$_2$NH$_2$ or for 3H$_2$NCH$_2$CH$_2$NH$_2$?

2. What precautions must be taken knowing that Ni(en)$_3$Cl$_2$ is soluble in water?

3. If you were to use a 32.5% ethylenediamine solution and the density of the solution is 0.985 g/mL, what is the mass of the ethylenediamine in 3.4 mL of this solution?

4. What is the function of the acetone in this experiment?

5. What are the safety precautions in the use of acetone?

EXPERIMENT 5: SYNTHESIS OF Ni(en)₃Cl₂

Post-Lab Questions: (Answers must be in full sentences and typed.)

1. The procedure calls for using between 0.500 to 0.650 g of the hydrate in the synthesis. Would a student using 0.500 g of hydrate obtain a smaller percent yield of TRIS than a student using 0.650 g? Explain your answer.

2. At the part when all of the limiting reactant has reacted, how many moles of the other reactant remain unreacted? Show your calculations & explain your answer clearly.

3. Based on the number of moles of en you used, how many moles of hydrate would you need for it to totally react? How would this information tell you which reactant is the limiting reactant? Show your work and explain clearly.

4. The liquid that goes through the filter paper into the filter flask is called the "filtrate." What <u>exactly</u> is in your filtrate? List all the substances that you can think of that is in the filtrate.

Experiment 6: CHEMICAL REACTIONS

Purpose: Observe different types of chemical reactions and write molecular, total ionic and net ionic equations for the reactions

Performance Goals:
- Describe observations made on reactions taking place
- Make observations of single replacement reactions of selected metals and place them in order of reactivity
- Apply the aqueous solubility rules to determine the identity of the precipitate in a double replacement reaction
- Classify the observed reactions according to the types of reactions (single replacement, double replacement, gas evolution and other oxidation-reduction reactions) in order to predict the identities of the products of the reaction
- Write molecular, total ionic and net ionic equations
- Identify spectator ions in a reaction

Introduction:

It is often very exciting to conduct chemical reactions in the laboratory. This is because we can see something happening before our eyes. Observation of chemical reactions and the proper description of these changes should be important concerns to the beginning student in chemistry. There are several clues that indicate when a chemical reaction has likely occurred. Some of the most common of these clues are shown:

- Formation of a gas (effervescence) without supplying an external source of heat
- Formation of an insoluble solid (particles form a cloudy suspension, which sometimes settle rapidly to the bottom)
- A change in color
- Disappearance of a solid
- Evolution or absorption of heat (reaction mixture gets warm or cold)

One has to be careful, however, because the clues are not always definitive of a chemical reaction. For example, formation of a gas could be due to a liquid vaporizing, which would be a physical change rather than a chemical reaction. Furthermore, an absence of these clues does not mean a chemical reaction did not occur. Such is the case when there is no visual difference between reactants and products. One must therefore learn to predict what the products might be based on the reactants and see whether or not our predictions match what is observed. Being able to categorize the type of reaction helps us predict the identity of the products.

Single Replacement Reaction: $A + BC \longrightarrow AC + B$

When an element (A) reacts with a compound (BC) by replacing a less active element (B) in the compound, the reaction is called a *single replacement reaction*. If the element is a metal, it loses electron(s) and becomes a cation. The element being displaced gains electron(s) and is transformed into a *free element*. For example, when a piece of aluminum foil is placed into an aqueous hydrochloric acid solution, Al replaces the H⁺ in the HCl, and H⁺ changes into its free state, $H_2(g)$.

$$2Al(s) + 6HCl(aq) \longrightarrow 2AlCl_3(aq) + 3H_2(g)$$

Table 6.1 shows a section of what is commonly called the *Activity Series* where metals are placed in order of increasing reactivity. Note that hydrogen is included even though it is not a metal because it is often used as a reference point. We often speak of a metal as being more or less active than hydrogen.

Table 6.1: Activity Series of Some Common Metals (and Hydrogen) in Aqueous Solution

Activity	Metal	Ions and electrons produced in the reaction
Least active	gold	$Au \longrightarrow Au^{3+} + 3\,e^-$
	platinum	$Pt \longrightarrow Pt^{2+} + 2\,e^-$
	silver	$Ag \longrightarrow Ag^+ + e^-$
	hydrogen	$H_2 \longrightarrow 2H^+ + 2\,e^-$
	lead	$Pb \longrightarrow Pb^{2+} + 2\,e^-$
	iron	$Fe \longrightarrow Fe^{2+} + 2\,e^-$
	aluminum	$Al \longrightarrow Al^{3+} + 3\,e^-$
	sodium	$Na \longrightarrow Na^+ + e^-$
Most active	potassium	$K \longrightarrow K^+ + e^-$

Aluminum is expected to react with HCl based on this activity series because aluminum is more active than hydrogen and therefore able to replace H in HCl in the equation. Effervescence would be due to the formation of the hydrogen gas.

Based on this activity series, we can make the following predictions:

$Ag(s) + HCl(aq) \longrightarrow$ no reaction   Ag is less active than hydrogen.
$2Na(s) + 2HCl(aq) \longrightarrow 2NaCl + H_2(g)$ Na is more active than hydrogen

$Al(s) + FeCl_2(aq) \longrightarrow AlCl_3(aq) + Fe(s)$ Al is more active than iron
$Al(s) + NaCl(aq) \longrightarrow$ no reaction   Al is less active than sodium

Double Replacement (or Metathesis) Reaction: $AB + CD \longrightarrow AD + CB$
This is a reaction when two compounds (AB and CD) exchange ions. A reaction occurs if a precipitate or a molecular compound is formed. Under this category there are three common sub-categories: Precipitation, Acid-Base and Gas-Evolution Reactions

Precipitation Reaction: When two aqueous solutions are mixed and a solid forms, the solid is called a *precipitate*, and the reaction is referred to as a *precipitation reaction*. For example, if one mixes an aqueous silver nitrate solution, $AgNO_3(aq)$, with an aqueous magnesium chloride solution, $MgCl_2(aq)$, a white solid appears. Initially it appears as a suspension and the mixture looks *cloudy* or *foggy*. Allowed enough time, the precipitate usually settles to the bottom of the mixture. The solution above the precipitate is called the *supernatant solution*. The new substance shows up as a precipitate because it is insoluble in water. Substances that are soluble in water remains dissolved in the supernatant solution.

Table 6.2: General Solubility Rules* for Ionic Compounds

You are expected to memorize these rules for your CHEM 121 class and for your lab final.

Note: **In general, all ionic compounds are solids to begin with**. Molecular compounds, however, may be solids, liquids or gases.
"**Soluble**" means the compound will dissolve in water.
"**Insoluble**" means it will <u>not</u> dissolve in water.

1. **Always soluble:**
   Salts containing cations Group IA cations and NH_4^+
   Salts containing anions NO_3^-, $CH_3CO_2^-$ (or $C_2H_3O_2^-$), ClO_4^-, ClO_3^-

2. **Mostly soluble, with exceptions:**
   Salts containing halides (Cl^-, Br^-, I^-) are soluble, except salts of **Ag^+, Pb^{2+}**
   Salts containing sulfates (SO_4^{2-}) are soluble, except salts of **Sr^{2+}, Ba^{2+}, Pb^{2+}**

3. **Mostly insoluble, with exceptions:**
   Salts containing hydroxides (OH^-) are insoluble, except salts of **cations in Group IA, NH_4^+, Ca^{2+}, Sr^{2+}, Ba^{2+}**

4. **Assume to be insoluble:**
   All other salts may be ***assumed*** to be insoluble. This would be your best guess. If experimental evidence is contrary to this rule, or if you were told otherwise, do not be surprised. You will not be held responsible for cases not covered by these rules

*You will find slightly different solubility rules from one textbook to another. Presented here are the most common ones that you will come across in General Chemistry I & II.

To determine the identity of the white precipitate, first predict the possible products of the mixture of $AgNO_3$(aq) and $MgCl_2$(aq). Since both $AgNO_3$ and $MgCl_2$ are ionic solids that dissolve in water to give Ag^+, NO_3^-, Mg^{2+}, and Cl^- ions, the product must be a new substance containing some combination of these ions. The predicted products would be $Mg(NO_3)_2$ and $AgCl$. To determine which one of the two possible products will precipitate out, examine the solubility table (Table 6.2). The information tells us $Mg(NO_3)_2$ is soluble in water, so it cannot be the precipitate observed. $AgCl$, on the other hand, is insoluble in water, and so we can conclude that the white precipitate must be $AgCl$.

EXPERIMENT 6: CHEMICAL REACTIONS

This chemical reaction is indicated by the ***molecular equation*** shown below:

$$2AgNO_3(aq) + MgCl_2(aq) \longrightarrow 2AgCl(s) + Mg(NO_3)_2(aq)$$

The ***total ionic equation*** is written by showing all water soluble ionic compounds in their ionic form:

$$2Ag^+(aq) + 2NO_3^-(aq) + Mg^{2+}(aq) + 2Cl^-(aq) \longrightarrow 2AgCl(s) + Mg^{2+}(aq) + 2NO_3^-(aq)$$

Remember products that are solids *(s)*, liquids *(l)* or gases *(g)* never break apart. Note that both the Mg^{2+} and NO_3^- ions appear on both sides of the ionic equation. Since these ions are not involved in forming the product, they are called ***spectator ions*** and can be canceled out by subtracting them from both sides of the equation.

$$2Ag^+(aq) + \cancel{2NO_3^-(aq)} + \cancel{Mg^{2+}(aq)} + 2Cl^-(aq) \longrightarrow 2AgCl(s) + \cancel{Mg^{2+}(aq)} + \cancel{2NO_3^-(aq)}$$

The resulting equation, with coefficients reduced to the lowest ratio, is called the ***net ionic equation***:

$$Ag^+(aq) + Cl^-(aq) \longrightarrow AgCl(s)$$

Whether a reaction is expected to be observed or not is based on the solubility of the products. If both products are soluble, no precipitate would be expected and no reaction will be observed. (The reaction is not based on the activity series discussed for single replacement reactions.)

Acid-Base Reaction: $\underline{A}B + C\underline{D} \longrightarrow \underline{A}\underline{D} + CB$

$\underline{H}B + C(\underline{OH}) \longrightarrow \underline{H(OH)} + CB$

       acid   base   water   salt

The reaction of an acid and a base containing hydroxide is simply a double replacement reaction where A is hydrogen ion and D is the hydroxide ion (OH^-). The product HOH is water (H_2O), a molecular compound.

An acid is generally recognized by the H that appears at the front of its formula (such as HCl, HNO_3, H_2SO_4). In this discussion we will confine our discussions to only bases that contain hydroxides.

The reaction of hydrochloric acid with sodium hydroxide is one such reaction:

$$HCl(aq) + NaOH(aq) \longrightarrow H_2O(l) + NaCl(aq)$$

Formation of a molecular compound is the driving force for this reaction. We will study acid-base reactions in another experiment.

Double Replacement Reaction that Produces a Gas

There are double replacement reactions where the product immediately decomposes and one of the product is a gas. For example, hydrochloric acid reacts with sodium hydrogen carbonate (commonly known as sodium bicarbonate) to initially form carbonic acid, which immediately decomposes to carbon dioxide and water:

$$HCl(aq) + NaHCO_3(aq) \longrightarrow H_2CO_3(aq) + NaCl(aq)$$
$$\downarrow$$
$$H_2O(l) + CO_2(g)$$

Overall reaction:
$$HCl(aq) + NaHCO_3(aq) \longrightarrow H_2O(l) + CO_2(g) + NaCl(aq)$$

Effervescence is evidence of the reaction taking place.

Oxidation-Reduction Reaction (Redox Reaction):
A single replacement can also be classified as an *oxidation-reduction* (*redox* in short) reaction. The free element is *oxidized* as it loses electron(s), and its oxidation number is increased. The cation it replaces is *reduced* as it gains electron(s), and its oxidation number is decreased. A redox reaction must always involve a change in oxidation number of usually two or more elements.

In the previous example of Al reacting with hydrochloric acid, each Al is oxidized as it loses three electrons, and is oxidation number increases from zero to +3. The cation it is replacing (H$^+$ in HCl) is reduced as it gains one electron each, and its oxidation number decreases from +1 to zero.

$$2Al(s) + 6HCl(aq) \longrightarrow 2AlCl_3(aq) + 3H_2(g)$$

Al^0	$\longrightarrow$	Al^{3+} + 3 e$^-$	(oxidation)
2e$^-$ + 2H$^+$(aq)	$\longrightarrow$	H$_2$(g)	(reduction)

In the balanced equation, there are two Al, and therefore a total of 6 e$^-$ is formed. There are six H$^+$ and a total of 6 e$^-$ that is added. The number of electrons lost must equal the number of electrons gained.

The reaction of Al with HCl is classified both as a single replacement reaction as well as a redox reaction, but there are redox reactions that are not single replacement reactions. For example, when one lights a gas burner, methane in natural gas reacts with molecular oxygen (O$_2$) in the air generating CO$_2$ and H$_2$O:

$$\overset{-4}{CH_4}(g) + 2\overset{0}{O_2}(g) \longrightarrow \overset{+4}{CO_2}(g) + 2\overset{-2}{H_2O}(l)$$

(loses 8 e$^-$ per C)
(gains 2 e$^-$ per O)
total of 8 e$^-$ transferred

Oxidation number of carbon in methane increases from –4 to +4 as it loses 8 electrons per carbon atom. At the same time, oxidation number of oxygen decreases from zero to –2. It is gaining 2 electrons per oxygen atom. In the balanced equation, there are 4 oxygen atoms. Therefore the total number of electrons being transferred is 8 electrons (from C to O).

To identify a reaction as a redox reaction, we only need to identify at least one element having changed its oxidation number.

Equipment/Materials:

Part I: Well plate, aqueous solutions in dropper bottles: 6 M HCl in dropper bottle, Cu, Zn and Mg (about 0.5 cm x 0.5 cm), disposable plastic pipets, toothpicks to act as stirrers, Q-Tips® for cleaning

Part II: Well plate, Q-tips® for cleaning well plate, toothpicks as stirrers, aqueous solutions in dropper bottles: 0.3 M HCl, 0.3 M Na_2CO_3, 0.3 M KOH, 0.1 M $Pb(NO_3)_2$, 0.1 M NaI, 0.1 M NaCl, 0.1M $Cu(NO_3)_2$, phenolphthalein indicator, (Demonstration: solid Na_2CO_3, vinegar, 0.1 M $Mg(NO_3)_2$, 0.1 M KOH, Mg strip about 2-3 cm long, gas burner, crucible tongs, Cu wire about 8-10 cm long)

SAFETY PRECAUTIONS
- Wear safety goggles until everyone has finished the reactions.
- Wash your hands thoroughly after the experiment and before leaving the lab. **Almost all of the chemicals in this experiment are toxic, irritant or corrosive**. Instead of watching out for only some of the chemicals, it would be safer if you just treated all of the chemicals with care. Avoid contact with your eyes, skin and clothing. Wash your hands immediately if you get any on your hands.
- $AgNO_3$ will darken your skin and the spots will remain on your skin until your skin wears off (in a couple of days).

Procedure:

Work with one partner in assembling the necessary chemicals for each reaction. Be sure both partners are present before mixing the chemicals together. Both partners must be making the observations together.

You will record and write equations for the entire experiment on the pages provided here. You will not be using your lab notebook for this experiment. You may also use pencil throughout this experiment.

These reactions will be performed on a small scale using plastic mini-pipets and a well plate. Begin by cleaning the well plate **THOROUGHLY** with the help of Q-tips® under running tap water. Then rinse it well with deionized water, and then shake off any remaining water. You do not need to dry it any further. You will be held responsible for incorrect results due unclean equipment.

Part I: Reactivity of Selected Metals

1. Place 10 drops of 6 M HCl in each of three wells side by side.

EXPERIMENT 6: CHEMICAL REACTIONS   77

2. Add a small piece of copper in the first well, a piece of zinc in the second, and a piece of magnesium in the third. Record your observations on the Report Sheet of this manual, including how fast the reaction is occurring: fast, medium fast, slow, no reaction.
3. On the Report Sheet, record which metal is the most reactive, which metal is the least reactive and which metal(s) is/are more reactive than hydrogen.
4. Based on your conclusions, place Cu, Zn and Mg in the Activity Series on the Report Sheet in the format shown in Table 6.1. Would you expect Mg to react with $Cu(NO_3)_2$?
5. In a separate well, place 10 drops of 0.1 M $Cu(NO_3)_2$ solution. Add a small piece of Mg. Record your observations. If there is no reaction, wait 5 minutes and check again.
6. Write molecular, total ionic and net ionic equations for the reactions specified on the Report Sheet.

Part II: Observations & Equations for Some Chemical Reactions

You will be performing five reactions. Four more reactions will be done as a demonstration; two during the pre-lab and two during the lab. For each reaction begin by describing the appearance of each reactant (its physical state, color and general appearance). Mix the two reactants in the well plate, record your observations and write the molecular, total ionic and net ionic equation *REGARDLESS OF WHETHER A REACTION IS OBSERVED*. Finally, classify each reaction by selecting one of the following:
- single replacement
- double replacement – precipitation
- double replacement – acid-base reaction
- double replacement – gas evolution reaction
- redox other than single replacement

A disposable graduated pipet is to be used to deliver 1.0 mL of each reactant solution to the well plate. Careful observation will determine whether a chemical reaction occurred.

You will have to observe very carefully to detect the small amount of precipitate or low volume of gas that forms. The mixture should be stirred well with a toothpick. The well plate should be placed on a dark background to detect white precipitate formation. Some reactions are fast and must be observed immediately after mixing. Others may be quite slow. If no reaction is detected, check the reaction mixture after 10 minutes.

In one mixture, the phenolphthalein indicator is used. Phenolphthalein is a chemical that is colorless when in an acidic environment, and bright pink when in a basic one. A sample containing this indicator changing from colorless to pink would indicate that a chemical reaction has taken place, converting it from acidic to basic. Note that phenolphthalein is not part of the reaction and is not be included in the equation.

Cleaning Up: The solutions in the well plate should NOT be poured down the sink, but instead they should be poured in the designated container in the fume hood. The well plate should then be washed thoroughly with tap water at the sink to remove any remaining residue. Make use of the Q-tips® provided to dislodge any residues that may be adhering to the well.

EXPERIMENT 6: CHEMICAL REACTIONS

Reaction Mixtures:

Reaction #1   0.3 M sodium carbonate and 0.3 M hydrochloric acid
Reaction #2   0.3 M hydrochloric acid, 2 drops of phenolphthalein indicator & record the color.
              Next add 0.3 M potassium hydroxide dropwise until a color change occurs.
Reaction #3   0.1 M lead(II) nitrate and 0.1 M sodium iodide
Reaction #4   0.1 M sodium chloride and 0.3 M sodium carbonate
Reaction #5   0.1 M copper(II) nitrate and 0.3 M potassium hydroxide
Reaction #6   (done as a demonstration by instructor) magnesium strip is ignited.
Reaction #7   (done as a demonstration by instructor) copper wire and 0.2 M silver nitrate

Pre-Lab Exercise:

In writing *total ionic equations*, only certain substances, such as NaCl *(aq)*, are to be converted into separate ions, Na^+ *(aq)* + Cl^- *(aq)*. The rule is very simple. A formula is converted into cations and anions IF AND ONLY IF:
   1) Substance has the physical state *(aq)*
   and 2) it is either an ***ionic compound*** or a ***strong acid***.
There are 7 common strong acids that you should commit to memory:
   HNO_3 H_2SO_4 $HClO_4$ $HClO_3$ HCl   HBr   HI
You can assume all other acids are weak acids.

You should be able to answer the following questions:
1. For each of the following substances, complete the equations to show whether it should be converted into separate ions. The first two have been done for you as examples:
   1. $Na_2SO_4(aq)$ $\longrightarrow$ $2Na^+(aq)$ + $SO_4^{2-}(aq)$
   2. $CO_2(g)$ $\longrightarrow$ $CO_2(g)$ (CO_2 does not form ions.)
   3. $K_3PO_4(aq)$ $\longrightarrow$
   4. $3Mg(NO_3)_2(aq)$ $\longrightarrow$
   5. $HBr(aq)$ $\longrightarrow$
   6. $NaHCO_3(aq)$ $\longrightarrow$
   7. $CH_3OH(aq)$ $\longrightarrow$
   8. $MgCl_2(s)$ $\longrightarrow$
   9. $H_2O(l)$ $\longrightarrow$
   10. $2NH_4Cl(aq)$ $\longrightarrow$

2. What precautions must be taken when handling 0.3 M HCl and 0.3 M KOH solutions in the laboratory?

3. What is the function of the phenolphthalein indicator?

4. When an aqueous solution of lead(II) nitrate is mixed with an aqueous solution of sodium iodide a yellow precipitate forms. Use the data in the solubility table to write the (1) molecular, (2) total ionic and (3) net ionic equations for this chemical reaction.

Experiment 7: ACID-BASE TITRATION: STANDARDIZATION OF A SOLUTION

Purpose: Determine molarity of a solution of unknown concentration by performing acid-base titrations

Performance Goals:
- Apply the concepts of titration and standardization
- Gain practice in the use of the analytical balance and buret
- Use acid-base titration to standardize a NaOH solution
- Calculate molar concentration of a NaOH solution

Introduction:

Titration is an analytical technique for determining the concentration of a solution (*analyte*) by measuring its volume required to completely react with a standard, which could be a solid of high purity or a solution of known concentration. The concentration of the analyte is calculated based on the stoichiometry of the reaction between the analyte and the standard. Different types of titrations such as acid-base titrations and oxidation-reduction titrations have been utilized in chemical analyses. In this experiment an acid-base titration will be used to determine the molar concentration of a sodium hydroxide (NaOH) solution. Acid-base titrations are also called neutralization titrations because the acid reacts with the base to produce salt and water.

During an acid-base titration, there is a point when the number of moles of acid (H^+ ions) equals the number of moles of base (OH^- ions). This is known as the *equivalence point*. For example, in the reaction between HCl and NaOH (Equation 1), the number of moles of H^+ will be same as the number of moles OH^- at the equivalence point since the molar ratio between HCl and NaOH is one-to-one. For a reaction between a diprotic acid such as sulfuric acid (H_2SO_4) that contains two moles of H^+ ions per mole of H_2SO_4 and a base such as NaOH that contains one mole of OH^- ions per mole of NaOH, the moles of H^+ and OH^- will still be same at the equivalence point, but the moles of acid (H_2SO_4) and base (NaOH) will not be same. There will be twice as many moles of NaOH because every mole of H_2SO_4 generates two moles of H^+ ions, requiring two moles of NaOH to neutralize them (Equation 2).

$$HCl(aq) + NaOH(aq) \longrightarrow NaCl(aq) + H_2O(l) \qquad \text{Equation 1}$$
$$H_2SO_4(aq) + 2NaOH(aq) \longrightarrow Na_2SO_4(aq) + 2H_2O(l) \qquad \text{Equation 2}$$

A chemical substance called an *indicator* is used to determine the equivalence point. An indicator is a solution that changes color based on the nature (acidity or basicity) of the solution. For acid-base titrations the indicators used are weak acids or bases. They have one color in acidic solutions and another color in basic solutions because they undertake different forms in different solutions. An acid-base indicator in acid form undergoes a change when put in water as illustrated by Equation 3.

$$HIn(aq) + H_2O(l) \rightleftharpoons In^-(aq) + H_3O^+(aq) \qquad \text{Equation 3}$$
acid form $\qquad\qquad\qquad$ base form

EXPERIMENT 7: ACID-BASE TITRATION: STANDARDIZATION

The indicator, phenolphthalein, is often utilized when strong acids and/or bases are used in a titration. Phenolphthalein is colorless in its acid form, but is pink in the presence of excess base. When phenolphthalein changes from colorless to a pale pink color as a base is added to an acid solution, it is an indication that the mixture has just passed the equivalence point. This observable instant is referred to as an *endpoint*. An indicator is chosen such that the observable *endpoint* occurs at or very close to the stoichiometric *equivalence point*. If the color is a bright pink, it is an indication that too much base has been added and we no longer have the equivalence point.

Solutions of strong bases, such as NaOH, are usually made by dissolving the solid in water. Presumably its concentration can be calculated from the mass of the solid and the volume of the solution produce. However, the solid NaOH is hygroscopic (easily absorb moisture from air.) In addition, the carbon dioxide that is naturally dissolved in water further changes the acidity of the solution. Thus, the actual concentration of the NaOH solution has to be determined experimentally using a process called *standardization*. To standardize a base solution such as NaOH, an acid whose amount can be determined to a high degree of accuracy (called a primary standard) is needed. Potassium hydrogen phthalate (KHP), a monoprotic acid, is often used as a primary standard for titrating bases. KHP can be dried in an oven to remove traces of water. The reaction between sodium hydroxide and KHP solutions proceeds according to the following reaction (Equation 4).

$$KHC_8H_4O_4$$

KHP + NaOH → Na-KHP + H$_2$O   Equation 4

KHP (MM = 204.2 g/mol)

Note the stoichiometry: one mole of KHP reacts with one mole of NaOH.

Alternatively, solutions of HCl with known concentration can be purchased commercially and used to standardize basic solutions. Such solutions are referred to as *standard solutions*. The reaction between solutions of HCl and NaOH is illustrated by Equation 1.

In this experiment, standardization of a NaOH solution will be carried out either using KHP as the primary standard or by using a standard HCl solution of known concentration.

Equipment/Materials:
Using KHP as standard: Buret, 250-mL Erlenmeyer flasks (three), ring stand, clamp, phenolphthalein, NaOH solution of unknown concentration, solid potassium hydrogen phthalate (KHP), electronic balance (magnetic stir bar and stir plate maybe supplied)

Using the HCl as standard: Burets (two), 250-mL Erlenmeyer flasks (three), ring stand, clamp, phenolphthalein, NaOH solution of unknown concentration, standard 0.1000 M HCl solution, (magnetic stir bar and stir plate may be supplied)

EXPERIMENT 7: ACID-BASE TITRATION: STANDARDIZATION   89

How to Record Buret Readings

1. The curved surface of a liquid is called a ***meniscus***. Water has a meniscus that curves down. It is important that you read the buret **_with the meniscus at eye level_** to avoid the problems of parallax. You know you are not at eye level if the markings on the buret on the backside are visible, as in Fig.1. If you are at eye level, the markings should be as in Fig. 2.

Fig. 1   Fig. 2

2. As usual, you record to 1/10 of the smallest increment of your measuring device. In the case of burets (see sketches below), what is the smallest division shown? How many decimal places should you record?

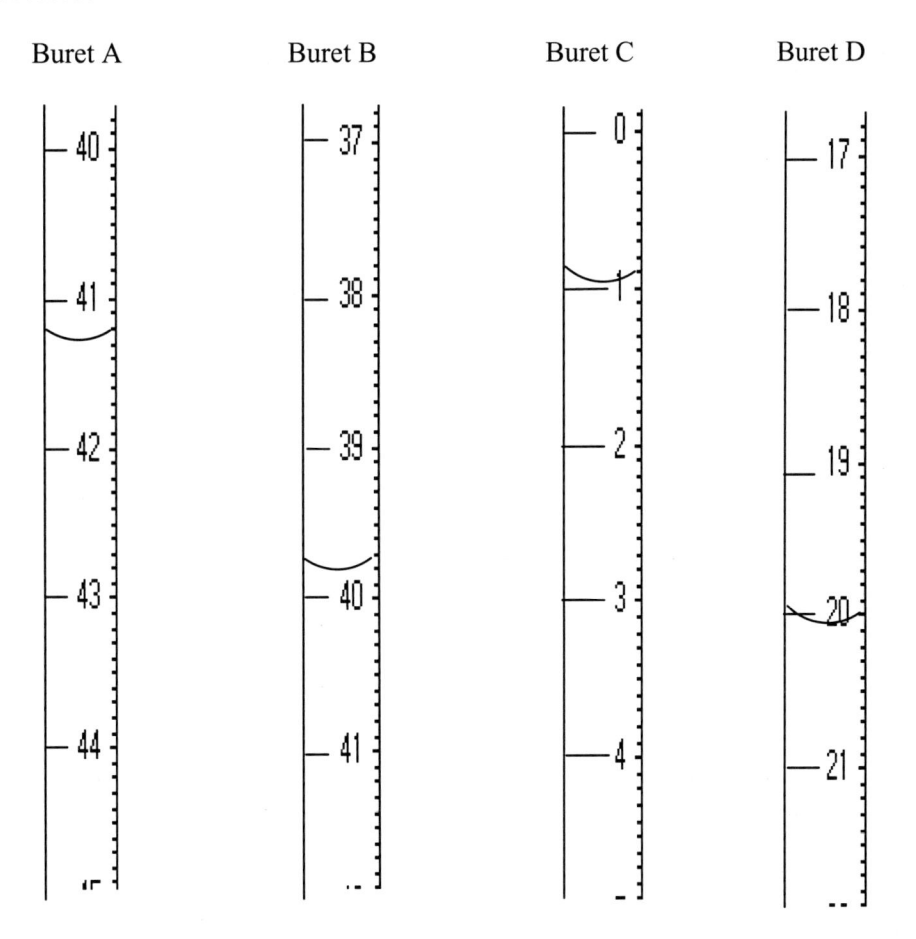

Buret A   Buret B   Buret C   Buret D

Buret Readings: A = 41.26 mL   B = _____   C = _____   D = _____

If the meniscus is "exactly" on 40 mL, how should you record it? Ans. _____

Examine the buret reading for A and then write down your readings for B through D. Your instructor will go over what those readings should be in the Pre-Lab discussion.

Procedure:

Cleaning the Buret: Having a clean buret plays an important role in the standardization process. Therefore, it is necessary to make sure that the buret is clean. Aqueous solutions and water in the buret should form a meniscus and should not form beads on the walls of the buret.

Your instructor or lab technician will indicate whether the burets supplied to you are clean. If they are not, follow the instructions below on how to clean them.

In order to clean a buret the stopcock of the buret should be turned perpendicular to the buret. Deionized or distilled water should be run down the sides of the buret using a wash bottle. Water should be poured out of the buret, rotating it so that water runs along the entire inner surface. Caution must be exercised not to hit the buret on the sink or any other objects. The above process should be repeated two more times. The buret should be filled with water. If a meniscus forms and water does not form beads on the sides, the buret is clean. If the buret is not clean, then the buret should be cleaned with soap and water followed by washing with deionized water three times. Make sure that no residue of soap remains inside the buret. The stopcock and the tip need to be cleaned by opening the stopcock and letting water flow through the stopcock and tip into a beaker.

Setting up the buret: Setting up the buret correctly is important for a successful acid-base titration. Figure 7.1 illustrates correct set up of a buret.

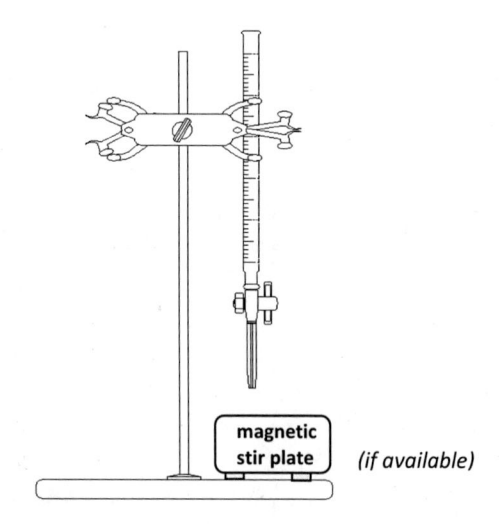

Figure 7.1. Correct setup of a buret

Check that the buret is vertical. If it is even slightly tilted, it means it has not been properly fastened to the buret clamp, and there is a danger of the buret popping off the clamp and splattering its content on you.

If a standard HCl solution is used, there would be a second buret fastened to the other side of the buret clamp.

EXPERIMENT 7: ACID-BASE TITRATION: STANDARDIZATION

Standardizing the NaOH Solution
In the lab notebook, set up a data table similar to the one given at the end of this exercise. Record all data directly into the data table in your lab notebook.
Your instructor will indicate which procedure you are to follow (Part I or Part II).

Some labs are equipped with magnetic stir bars and stir plates. If none are available, your instructor will demonstrate how to swirl the contents of the Erlenmeyer flask.

Part I: Using KHP for Standardization
1. Label a 250-mL beaker as "NaOH" and a 400-mL beaker as "Waste."
2. Set up a ring stand with a buret clamp and place a clean buret in the buret clamp. Position a magnetic stir plate in such a manner that the buret tip is directly over the center of the stir plate as shown in Figure 7.1.
3. Place the waste beaker under the buret and close the stopcock.
4. Pour about 100 mL of the NaOH solution into the NaOH beaker. Rinse the buret with the NaOH solution twice with approximately 10-mL portions. The rinse solution should go into the waste beaker. Fill the buret to the very top. Next drain about 5 mL of the solution through the tip to rinse and fill the tip with the solution. *(Hint: If you tilt the buret slightly, it will make pouring much easier. The use of a funnel is not recommended, but if you decide to use one, it must be absolutely clean and dry, and you must remember to remove it from the buret before you start your titration.)* Your instructor will demonstrate how to rinse a buret properly.
5. At the end of rinsing, check to make sure there are no air bubbles at the tip of each buret. Next, carefully drain enough NaOH solution from the buret so that the meniscus is on the zero calibration mark or slightly below.
6. Record the actual volume reading on the buret. This will be called the initial reading of the buret. It does not have to be at zero. Remember that the volume markings on burets are read from top to bottom and volumes are recorded to two decimal places. The first decimal place can be read accurately and the second decimal place needs to be estimated.
7. Label three clean 250-mL Erlenmeyer flasks (labeled 1, 2, and 3). They do not need to be dry.
8. Tare a plastic weighing boat to zero. Add about 0.3–0.4 g of KHP and record the exact weight. Transfer the KHP sample carefully into Erlenmeyer flask #1. Repeat for flasks #2 and 3, being careful not to mix up the flasks.
9. Add about 50 mL of distilled water to each flask (use graduations on the flask) and 3 drops of phenolphthalein indicator to each flask. It may be necessary to warm the flasks slightly on a hot plate in order to dissolve the KHP. However, the solutions must cool back to room temperature before you begin the titration.
10. Slide a clean magnetic stir bar into the flask, taking care not to let any solution splash out.
11. Check the initial buret reading. If it has changed, your buret has a leak.
12. Turn on the magnetic stirrer and titrate the samples with NaOH solution from the buret until the first shade of pink persists for 30 seconds. As you near the endpoint, use the wash bottle of deionized water to wash down any splatter that may have collected on the inside surface of the flask.

EXPERIMENT 7: ACID-BASE TITRATION: STANDARDIZATION

13. Calculate the ratio of mass of KHP (g) to Volume of NaOH (mL) for each trial. The volume in each trial will differ because not the same mass of KHP is used. The ratio, however, should be the same. This calculation will indicate whether or not all three trials are acceptable and whether more trials are necessary. The ratios should not differ by more than 0.001 g KHP/mL NaOH. This calculation must be performed before proceeding to the next trial.

For example, if 0.346 g of KHP is used, the initial buret reading is 1.07 mL and the final buret reading is 32.82 mL, the volume of NaOH at the endpoint would be (32.82–1.07) mL = 31.75 mL. The ratio of mass of KHP to Volume of NaOH would be

$$\frac{0.346 \text{ g KHP}}{31.75 \text{ mL NaOH}} = 0.0109 \text{ g KHP/mL NaOH}$$

Sample Data Table for Standardization Using KHP

Trial #	#1	#2	#3
Mass of KHP (g)			
Final buret reading (mL)			
Initial buret reading (mL)		Sample Only	
Volume of NaOH used (mL)			
Ratio of Mass KHP/Volume of NaOH used (g KHP/mL NaOH)			

Calculations for Using KHP for Standardization

1. Calculate the # mol of KHP by using the mass of KHP used in each trial and the molar mass.

   $$\text{\# mol KHP} = \text{\# g of KHP} \left(\frac{1 \text{ mol KHP}}{\text{molar mass of KHP}} \right)$$

2. Find the # mol NaOH by using the relationship, #mol NaOH = #mol KHP
3. Find the volume of NaOH used in each trial by subtracting the initial buret reading from the final buret reading; V_{NaOH} = Final buret reading – Initial buret reading
4. Convert V_{NaOH} in mL to L
5. Calculate molarity of NaOH by using the equation

   $$\text{Molarity of NaOH} = \frac{\text{\# mol NaOH}}{\text{\# L NaOH soln}}$$

   Alternatively, molarity of NaOH can be calculated by using the following setup:

   $$\text{Molarity of NaOH} = \text{\#g KHP} \left(\frac{1 \text{ mol KHP}}{\text{molar mass KHP, g}} \right) \left(\frac{1 \text{ mol NaOH}}{1 \text{ mol KHP}} \right) \left(\frac{1}{\text{\# mL NaOH}} \right) \left(\frac{10^3 \text{ mL NaOH}}{1 \text{ L NaOH}} \right)$$

6. Calculate average molarity of NaOH.

Part II: Using HCl for Standardization

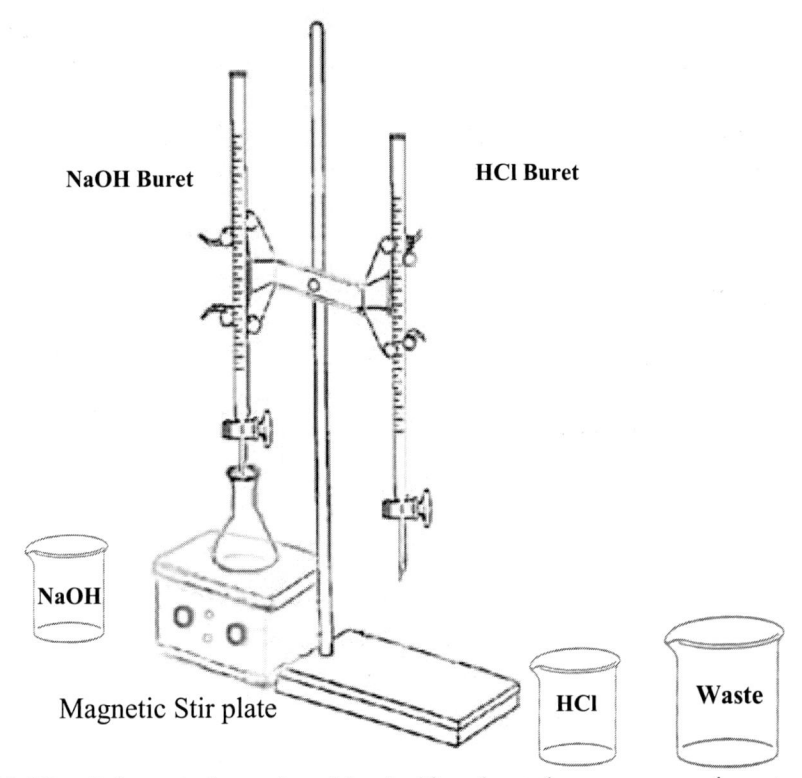

[Diagram of titration setup showing NaOH Buret and HCl Buret mounted on a ring stand, with an Erlenmeyer flask on a Magnetic Stir plate, and beakers labeled NaOH, HCl, and Waste.]

1. Obtain two 50.00-mL burets from the side shelf and set them up on a ring stand fitted with a buret clamp as shown in the diagram below. Place gummed labels provided on the <u>buret clamp</u> to label one buret as HCl and one as NaOH. Next, label a 250-mL beaker as HCl and a 250-mL beaker as NaOH. Both beakers must be clean and dry. Finally, label a 400-mL beaker as "Waste." It is essential that you keep all your equipment clearly marked.
2. Position a magnetic stir plate under the NaOH buret as shown in the figure above.
3. Pour about 100 mL of the NaOH solution into the NaOH beaker, and 100 mL of the standard 0.1000 M HCl solution into the HCl beaker. Record this concentration in your notebook. ***Do not get your containers mixed up***! Rinse each buret with the corresponding solution twice with approximately 10-mL portions and then fill the buret to the very top. Next drain about 5 mL of the solution through the tip to rinse and fill the tip with the solution. (Hint: If you tilt the buret slightly, it will make pouring much easier. The use of a funnel is not recommended, but if you decide to use one, it must be absolutely clean and dry, and you must remember to remove it from the buret before you start your titration. Obviously you cannot use the same funnel for the NaOH and the HCl solutions.) Your instructor will demonstrate how to rinse a buret properly. Remind him/her to do so if he/she forgets.
4. At the end of rinsing, check to make sure there are no air bubbles at the tip of each buret. Next, carefully drain enough NaOH solution from the buret so that the meniscus is on the zero calibration mark or slightly below. Do the same for the buret holding the HCl solution.
5. Obtain a 250-mL Erlenmeyer flask from your drawer. Clean it thoroughly with detergent using a test tube brush, and rinse thoroughly with tap water. Finally, rinse

it several times with small portions of deionized water. This flask does not need to be dried. Obtain a wash bottle of deionized water and have it next to you, ready for use.
6. Record the **initial** readings to 0.01 mL for **both** burets for Trial 1 in your lab notebook. Next, allow the HCl solution to drain into the Erlenmeyer flask until the meniscus is around 25.0 mL. If there are any drops hanging on the tip of the buret, touch it to the inside wall of the Erlenmeyer. You do not need to record this reading yet.
7. Carefully slide a magnetic stirring bar into the Erlenmeyer flask, taking care not to allow any solution to splash out.
8. Add one or two drops of phenolphthalein indicator to the flask, and place a white sheet of notebook paper under the flask to help you see the color change better. Turn the magnetic stirrer on gradually to medium speed. (Be careful you are not turning on the heat by mistake!)
9. Slowly add the NaOH solution from the buret to the flask drop by drop. Initially you will see a pink color appear and disappear as the NaOH gets neutralized by the HCl in the flask. As you near the endpoint, use the wash bottle of deionized water to wash down any splatter that may have collected on the inside surface of the flask.
10. When the pink color is staying longer, slow down the addition of the drops so that at the end point you have time to turn off the stopcock before the next drop is added. Continue adding the NaOH solution until the color changes to a pale pink that persists after 30 seconds of stirring. This should be a very pale pink; the paler, the better.
11. If it is darker than what your instructor has shown you, it means you have added too much base and have past the endpoint. Normally this means you have to throw out the solution and start over; however, since you have the acid in the other buret, all you have to do is add one or two drops of acid to the flask to remove the pink color, and you can try again to add *just* enough base to reach the pale pink. **After you have reached a satisfactory end point, record the final buret readings to the nearest 0.01 mL for *BOTH* burets.**
12. Calculate the net volumes (subtract initial buret reading from final buret reading) of NaOH solution, and net volumes of HCl for the three titrations. Then, calculate the ratio of the volume of HCl to volume of NaOH. These **ratios should not differ by more than 0.04**. If they do, additional titrations may be necessary before you go on. YOU MUST DO THESE CALCULATIONS BEFORE YOU DO THE NEXT TRIAL!
13. It is safe to discard the contents of the Erlenmeyer in the sink as it is near neutral. Take care that the stir bar does not go down the drain! Rinse the flask and the stir bar thoroughly with deionized water. Re-fill the burets with the corresponding solutions and repeat for Trials 2 and 3 (Steps 5 through 6). Do not forget to record the **initial** buret readings for **both** burets at the beginning and the **final** buret readings for **both** burets at the end point. Also do not forget to add the indicator!

EXPERIMENT 7: ACID-BASE TITRATION: STANDARDIZATION

Sample Data Table for Standardization Using HCl

Molarity of Standard HCl (from label on bottle) =			
Trial #	# 1	# 2	# 3
Final buret reading of HCl (mL)			
Initial buret reading of HCl (mL)			
Volume of HCl used (mL)		Sample Only	
Final buret reading of NaOH (mL)			
Initial buret reading of NaOH (mL)			
Volume of NaOH used (mL)			
Ratio of Volume HCl/Vol NaOH			

Calculations for Using HCl for Standardization

1. Calculate the number of moles of HCl by using the volume of HCl used in each trial and its molarity
   # mol HCl = Molarity of HCl (mol/L) x Volume of HCl (L)
2. Find the number of moles NaOH from the number of moles of HCl (See Equation 1).
3. Calculate molarity of NaOH of each trial by using the equation:
   Molarity = #mol NaOH/#L NaOH soln
4. Calculate the average molarity of NaOH.
5. Calculate the deviation of each trial, the average deviation and the relative average deviation (RAD). Refer to Experiment #1 for these calculations.

Pre-Lab Exercise:

1. What is a *primary standard* substance? Give an example.

2. What is meant by the term, *equivalence point*, during an acid-base titration?

3. What is an indicator? What role does it play in an acid-base titration?

4. Why is it necessary to standardize a NaOH solution?

5. Find the molar concentration of a NaOH solution if 0.3999 g of it is dissolved in water to yield 100.0 mL solution. Watch your sig. fig.

6. Consider the following equation
   $$KOH(aq) + HCl(aq) \rightarrow KCl(aq) + H_2O(l)$$
   Calculate the molar concentration of a KOH solution if 25.0 mL of it was required to completely neutralize 37.5 mL of a 0.0800 M HCl solution.

7. What is the criterion to decide whether you need to do more then three titrations?

Post-Lab Questions:

1. Using the average molarity of NaOH calculate % error in your determination of the molarity of NaOH if the true value of the molar concentration were 0.1021 M.

EXPERIMENT 7: ACID-BASE TITRATION: STANDARDIZATION

2. If your experimental molar concentration of NaOH were not close to the accepted value what are two likely reasons for the discrepancy. Do not simply say "human error." Explain clearly and specify whether the cause of error would lead to a higher or lower experimental molar concentration of NaOH.

3. A buret is graduated with zero at the top and 50.00 mL at the bottom. If the buret is filled with water and has an initial buret reading of 1.08 mL and 35.28 mL is removed from the tip, what would the final buret reading be?
1.08 + 35.28

4. The purpose of the experiment was to determine the concentration of a NaOH solution. During the titration more water was added to wash down the sides of the flask. This would change the concentration in the flask. Why is this not a concern? Explain your answer.

Experiment 8: ACID-BASE TITRATION: DETERMINATION OF THE EQUIVALENT WEIGHT OF AN UNKNOWN ACID

Purpose: Identify an unknown acid by determining its equivalent weight

Performance Goals:
- Gain more practice in performing acid-base titrations
- Determine the equivalent of an unknown acid
- Determine the identity of the unknown acid by comparing the experimentally determined equivalent weight with equivalent weights of some known acids

Introduction:

Equivalent weight (EW) is the weight in grams of one *equivalent* (eq). It has units of g/eq. In an acid-base reaction, an *equivalent* is the amount of a substance that produces or reacts with <u>one mole of hydrogen ions</u>, H^+. For an oxidation-reduction reaction, an equivalent is defined as the amount of substance that produces or reacts with <u>one mole of electrons</u>. In this experiment we will focus only on acid-base reactions.

For example, one mole of HCl is one equivalent because it can produce one mole of H^+. The mass of one mole of HCl would therefore be its equivalent weight. Since the mass of one mole of HCl is also its molar mass (MM), its equivalent weight and molar mass have the same numerical value. In an acid such as H_2SO_4, one mole can produce <u>two</u> moles of H^+. One mole of H_2SO_4 is therefore <u>two</u> equivalents, and one equivalent is ½ mole of H_2SO_4. Its equivalent weight would be the mass of ½ mole of H_2SO_4. It has a numerical value that is half of that of its molar mass.

For a base such as NaOH, it can react with one mole of H^+. Therefore one mole of NaOH is one equivalent. For $Ba(OH)_2$, however, one equivalent is ½ mol because it takes only half a mole to react with one mole of H^+.

$$NaOH(aq) + H^+(aq) \longrightarrow H_2O(l) + Na^+(aq)$$
$$Ba(OH)_2(aq) + 2H^+(aq) \longrightarrow H_2O(l) + Ba^{2+}(aq)$$

Equivalent weight and molar mass have units shown below:

$$EW = \frac{\# g}{\# eq} \quad \text{(Equation 1)} \qquad MM = \frac{\# g}{\# mol} \quad \text{(Equation 2)}$$

Equivalent weights may be calculated from molar masses if the chemistry of the substance is well known. If we know how many moles of H^+ (# of eq) are produced by one mole of acid, we can use it as a conversion factor of (# eq/mol acid) or (1 mol acid/# eq):

$$EW = \frac{MM \text{ g}}{1 \text{ mol}} \times \left(\frac{1 \text{ mol}}{\# eq} \right)$$

For example, one mole of sulfuric acid, H_2SO_4, releases two moles of hydrogen ions (2 eq). Its molar mass is 98.08 g/mol. Therefore, its equivalent weight has a value half of its molar mass.

$$EW = \frac{98.08 \text{ g } H_2SO_4}{1 \text{ mol } H_2SO_4} \times \left(\frac{1 \text{ mol } H_2SO_4}{2 \text{ eq}} \right) = \left(98.08 \times \frac{1}{2} \right) \text{ g/eq}$$

EXPERIMENT 8: ACID-BASE TITRATION: EQUIVALENT WEIGHT OF ACID

Normality is a concentration unit similar to *molarity*. Recall that *molarity* is the number of moles of solute per liter of solution (#mol/L soln). *Normality* is the number of equivalents per liter of solution (#eq/L soln).

$$\text{Normality} = \frac{\text{\# eq}}{\text{\# L soln}} \qquad \text{Equation 3}$$

Rearrangement of Equation 3 gives us the following:

$$\text{\# eq} = \text{Normality}\left(\frac{\text{eq}}{\text{L soln}}\right) \times (\text{\# L soln}) \qquad \text{Equation 4}$$

Acid-base titrations can be utilized to determine the equivalent weight of a substance. In this experiment, a solution of an unknown acid will be titrated with a previously standardized NaOH solution (Experiment #7). At the equivalence point, the number of equivalents of the acid (A) will equal the number of equivalents of the base (B). The number of equivalents of the base can be determined by applying Equation 4 from above.

$$\text{\# eq B} = N_B \, V_B \qquad \text{Equation 5}$$

where N_B = normality of the base in eq/L
V_B = volume of the base in L

At the equivalence point,
$$\text{\# eq A} = \text{\# eq B}$$

Substituting ($N_B \, V_B$) for # eq B from Equation 5 gives
$$\text{\# eq A} = N_B \, V_B$$

$$\text{Equivalent weight of A} = EW_A = \frac{\text{mass of A}}{\text{\# eq A}}$$

$$EW_A = \frac{\text{mass of A}}{N_B \, V_B}$$

The possible unknown acids in this experiment are given in the Table 8.1

Table 8.1: Names and Equivalent Weights of Unknowns

Name of Acid	Equivalent Weight (g/eq)
succinic acid	59.04
adipic acid	73.07
sulfamic acid	97.10
mandelic acid	152.1

Equipment/Materials:
50-mL buret, 250-mL Erlenmeyer flasks (three), unknown acid, phenolphthalein, standardized NaOH solution, ring stand, clamp, stir bar, stir plate

Procedure:
In your lab notebook, set up a data table similar to the one below. Record all data directly into the data table in your lab notebook.

> Some labs are equipped with magnetic stir bars and stir plates. If none are available, your instructor will demonstrate how to swirl the contents of the Erlenmeyer flask.

EXPERIMENT 8: ACID-BASE TITRATION: EQUIVALENT WEIGHT OF ACID

1. Set up a 50-mL buret as in Experiment #7. Be sure to clean the buret (if it is dirty). Rinse it with approximately 10 mL of the standardized NaOH solution (from Experiment #7). Fill the buret with the NaOH solution. Run some of the solution through the tip and take the initial reading of the buret to 2 decimal places.
2. Tare a weighing boat to zero, weigh out between 0.35 g and 0.40 g of the unknown acid and record the exact mass of the sample in your lab notebook. Carefully transfer all of unknown into a clean 250-mL Erlenmeyer flask. Slide a clean magnetic stir bar into the flask, taking care not to let any solution splash out. Add about 50 mL of distilled water to the flask (use graduations on the flask), stir and dissolve the acid completely. Add 3 drops of phenolphthalein indicator to the flask.
3. Turn on the magnetic stirrer. Open the valve of the buret and slowly add NaOH solution to the acid solution in the Erlenmeyer flask. Continue adding NaOH solution until the first shade of pink persists for 30 seconds. Take the final buret reading (to 2 decimal places).
4. Calculate the ratio of mass of the unknown sample to volume of NaOH for each trial. This calculation will indicate whether or not all three trials are acceptable and whether more trials are necessary. **The ratios should not differ by more than 0.001 g unknown/mL NaOH.** This calculation must be performed before proceeding to the next trial.
5. Repeat steps 2 – 4 two more times. Make sure that the buret has enough NaOH in it before starting each run.

Sample Data Table

Unknown Acid Code # =			
Normality of NaOH (from Expt #7) =			
	Trial #1	Trial #2	Trial #3
Mass of unknown acid sample (g)			
Final buret reading (mL)			
Initial buret reading (mL)			
Volume of NaOH used (mL)	Sample only		
Ratio of mass unknown/volume NaOH (g unk/mL NaOH) *Record in decimal form.*			

Calculations:

1. Find the volume of NaOH added in each trial by subtracting the initial buret reading from the final buret reading: V_{NaOH} = Final buret reading – Initial buret reading
2. Convert volume of NaOH from mL to L.
3. Find the normality of the NaOH solution. NaOH has one OH⁻ and so will react with one equivalent of H⁺ ion. Therefore, the numerical values of molarity and normality of a NaOH solution are same. The molarity was determined in Experiment 7.
4. Use the following equation to calculate the equivalent weight of the acid for each trial and then find the average equivalent weight.

$$EW_A = \frac{\text{mass acid}}{N_B \, V_B}$$

N_B is the normality of NaOH solution, and V_B is the volume (in liters) of NaOH needed to reach the endpoint for each trial.

5. Compare the average equivalent weight with the equivalent weight in Table 8.1 and identify the unknown acid.
6. Calculate the deviation, average deviation and percent average deviation (RAD). Refer to Experiment 1 for these calculations.
7. Calculate the error and percent error in your average equivalent weight.

Pre-Lab Exercise:

1. What is meant by "equivalent weight"? Predict the equivalent weight of H_3PO_4.

2. How many equivalents are there in one mole of oxalic acid, HOOC–COOH?

3. Define normality. What will be the normality of a solution of $Ca(OH)_2$ if its molarity is 2.00 M?

4. An unknown acid sample weighing 0.332 g was dissolved in water and then titrated with 0.08321 N NaOH. The initial buret reading for the NaOH was 0.15 mL and the final buret reading at the endpoint was 34.30 mL. What is the equivalent weight of the unknown acid?

Post-Lab Questions:
Answers must be typed and in full sentences.

1. What should you examine to determine the precision of your results? Explain.

2. What should you examine to determine the accuracy of your results? Explain.

3. If the standardization of the NaOH in Experiment 7 was not performed correctly and the experimental normality was too low, how would that affect the experimental equivalent weight being determined in this experiment? Would it be too high or too low? Explain.

4. You were allowed to weigh out between 0.35 g and 0.40 g of the unknown sample. If you used 0.35 g for Trial #1 and 0.40 g for Trial #2, would the experimental equivalent weight for Trial 2 be higher, lower or the same as that of Trial #1? Explain.

Experiment 9: CALORIMETRY

Purpose:
Part I: Identify an unknown metal by determining its specific heat
Part II: Determine the molar heat of neutralization of an acid-base reaction

Performance Goals:
- Determine specific heat of a metal
- Identify an unknown metal by its specific heat
- Calculate heat of reaction of an acid-base reaction
- Determine heat of a reaction using the heat of solution
- Calculate change in molar enthalpy of a neutralization by using heat of reaction

Introduction:

Thermochemistry is the study of the heat released or absorbed during the course of a physical or chemical transformation. Reactions (physical or chemical) that release heat are said to be *exothermic*, and those that absorb heat are said to be *endothermic*.

A *calorimeter* is the apparatus used in the measurement of the quantity of heat transferred during a reaction. The type used in this experiment is commonly called a "coffee cup calorimeter." It consists of a reaction chamber made of double-nested Styrofoam cups, a lid, and a temperature measuring device. The reaction takes place inside the cup. It is assumed that any heat lost by the *system* (the reactants) is totally transferred to the *surroundings* (calorimeter and the solution in it). Conversely, any heat produced by the system is totally gained by the surroundings. That is, we assume no heat is lost to the universe beyond the walls of the cup.

The letter, q, stands for the heat transferred in a reaction. If a reaction is exothermic, q has a negative sign, and if endothermic, it has a positive sign. In an exothermic reaction, q_{system} is negative because it is producing heat, but $q_{surroundings}$ is positive because the surrounding is absorbing the heat produced by the reaction. Note that the magnitudes of q_{system} and of $q_{surroundings}$, however, must be the same. They differ only in the sign. Thus we can write the equation

$$q_{system} = - q_{surroundings} \qquad \text{Equation 1}$$

Note that the minus sign does **NOT** mean q_{system} is negative. It merely means that the two q's must have opposite signs. For example, if a reaction absorbs 4 kJ, q_{system} would be +4 kJ, and $q_{surroundings}$ would have the same value, but the opposite sign, –4 kJ (supplying the heat that is absorbed by the system): $q_{surroundings} = -(+4 \text{ kJ}) = -4 \text{ kJ}$.

In this experiment, two reactions are studied, one physical and one chemical. The first involves the transfer of heat from a hot object to water at room temperature. The second involves a neutralization reaction between HCl and NaOH.

Specific Heat of an Unknown Metal

The specific heat (s) of a pure substance is defined as the amount of heat needed to raise the temperature of one gram of the substance by one degree (either Celsius or kelvin). It makes no difference whether the rise in temperature is in Celsius or kelvin but in this experiment °C will be used. Specific heat is a characteristic of a substance and can be used as supportive evidence to determine the identity of an unknown metal. By definition this will always be a positive number.

The relationship between heat and temperature change can be expressed by the equation shown below:

$$q = m \times s \times \Delta T \qquad \text{Equation 2}$$

where q = amount of heat transferred (in J)
m = mass of the substance (in g)
s = specific heat of the substance (in $J \cdot g^{-1} \cdot °C^{-1}$)
$\Delta T = T_{final} - T_{initial}$ (in °C)

Rearrangement of Equation 2 gives us the following:
$$s = \frac{q}{m \, \Delta T} \qquad \text{Equation 3}$$

In Part I of this experiment, the specific heat of a metal will be determined by heating a pre-weighed amount of a metal sample in a test tube and then dropping the hot metal into a coffee-cup calorimeter containing a measured amount of water at room temperature. In the process the temperature of the metal will decrease and the temperature of the water will increase. Heat exchange between the metal sample and water will stop after they reach the same temperature. The amount of heat lost by the metal sample (q_{metal}) will be equal to the amount of heat gained by the water (q_{water}) and the calorimeter ($q_{calorimeter}$).

In this part of the experiment, q_{system} is the same as q_{metal}, and $q_{surroundings}$ is a combination of $q_{calorimeter}$ and q_{water}. "Surroundings" consists of the calorimeter and the water within it. Thus we have the overall equation

$$q_{metal} = - [q_{calorimeter} + q_{water}] \qquad \text{Equation 4}$$

The Styrofoam cups (the calorimeter) have so little mass that they do not absorb a significant amount of heat. This simplifies our calculations because we can assume that the water in the calorimeter absorbs all of the heat from the reaction and the tiny amount absorbed by the calorimeter is insignificant. That is, $q_{calorimeter}$ = zero. The equation can therefore be simplified:

$$q_{metal} = - q_{water} \qquad \text{Equation 5}$$

As described previously, the magnitudes of the two q's are the same but they will have different signs since one will lose heat whereas the other will gain the same amount of heat.

Applying Equation 2 to the heat transfer that occurs in the water allows us to calculate the amount of heat that is transferred: $q_{water} = s \times m \times \Delta T$

where s = specific heat of water = 4.184 $J \cdot g^{-1} \cdot °C^{-1}$. The mass ($m$)

and change in temperature (ΔT) of the water are measured. We next apply Equation 5 to determine the heat transfer for the metal from the heat transfer for the water ($q_{metal} = -q_{water}$). Using Equation 3, the specific heat of the metal can be determined.

$$s = \frac{q}{m \, \Delta T}$$

where q = q_{metal}
m = mass of the metal
$\Delta T = (T_{final} - T_{initial})$ of the metal

By comparing the experimental specific heat to the specific heat values given below the identity of the unknown metal can be established.

Metal		Specific Heat $(J \cdot g^{-1} \cdot °C^{-1})$
lead	(Pb)	0.13
tungsten	(W)	0.13
tin	(Sn)	0.21
copper	(Cu)	0.39
zinc	(Zn)	0.39
aluminum	(Al)	0.91

Molar Heat of Neutralization (or Molar Enthalpy of Neutralization)

The amount of heat transferred during a chemical reaction is called the *heat of reaction,* an extensive property that is proportional to the amount of the limiting reactant used. The heat of reaction to be examined in Part II of this experiment is the *heat of neutralization* (the heat transferred during the reaction between an acid and a base). The term, ***molar** heat of reaction,* refers to the amount of heat transferred ***per mole*** of the specified reactant. It is, by definition, an intensive property. Since neutralization is always exothermic, the molar heat of neutralization of HCl therefore refers to the amount of heat produced by one mole of HCl as it reacts with a base (NaOH in this case).

$$HCl(aq) + NaOH(aq) \longrightarrow NaCl(aq) + H_2O(l) \qquad \text{Equation 6}$$

In a chemical reaction, the *reaction* is the *system*. Since this is a neutralization reaction, we will refer to the heat produced as q_{neutr}. The heat is absorbed by the calorimeter (assumed to be zero, as discussed earlier) and by the solution in the coffee cup calorimeter, which consists of HCl(aq) and NaOH(aq) at the beginning of the reaction and NaCl(aq) and water at the end of the reaction. We will refer to the heat absorbed by the solution as q_{soln}.

$$q_{neutr} = -q_{soln}$$

The value of q_{soln} will be determined by measuring the change in temperature (ΔT) of the solution inside the calorimeter during the reaction. Applying Equation 2 to the heat transfer of the solution

$$q_{soln} = m\ s\ \Delta T$$

where q_{soln} = amount of heat absorbed by the solution
   m = mass of the solution
    = mass of HCl soln + mass of NaOH soln
   s = specific heat of the solution = 4.18 J·g^{-1}·°C^{-1}
    (assume to be the same as the specific heat of water)
   ΔT = change in temperature of the solution = $T_{final} - T_{initial}$

In this experiment mass of the solution will be calculated using the volume of the solution and its density (assumed to be 0.997 g·mL^{-1})

Heat of neutralization is the same value as q_{soln} but with the opposite sign. The solution is absorbing heat so q_{soln} is positive. The reaction is exothermic, so q_{neutr} should be negative.

Relationship between q and ΔH: The term, q, refers to the heat flow measured under the conditions of the experiment. If the reaction takes place in an open vessel (not in a sealed container), the pressure is equal to the atmospheric pressure and we say that the q is under constant pressure (given the symbol q_p). Under this condition, q is equal to ΔH (enthalpy change of a reaction). Note that ΔH, like q, is an extensive property, proportional to the amount of limiting reactant used. The calculation described above was for the amount of heat transferred for a particular mass of reactants. The final step in the calculations would be to determine the <u>molar</u> heat of neutralization (or *molar ΔH, the molar enthalpy change of neutralization*). It is determined by dividing q_{neutr} by the number of moles.

Equipment/Materials:

Part I: Double-nested Styrofoam cups, two temperature probes, 50-mL beaker, 250-mL beaker, 400-mL beaker, large test tube, boiling chips, unknown metal, utility clamp, ring stand, hotplate, electronic balance

Part II: Double-nested Styrofoam cups, two temperature probes, two 250-mL beaker, two 50-mL graduated cylinders, large test tube, 1.00 M HCl, 1.00 M NaOH

Procedure: *Work with one partner but perform calculations individually.*

Part I: Specific Heat of an Unknown Metal
1. Place approximately 300 mL of tap water in a 400-mL beaker. Add two boiling chips and heat to boiling.
2. An unknown metal will be assigned to you and your partner. Record the unknown code number.
3. Record the mass and appearance of the metal. If you were given a metal cylinder, you can place it directly on the balance pan. If you were given metal pellets, tare a small beaker (50-mL beaker) to zero, remove the beaker from the pan, transfer all of the metal pellets into the beaker and record the mass in your lab notebook.
4. Gently slide the metal into a large test tube. Fasten a utility clamp near the mouth of the test tube. You must be able to safely hold the test tube up by the utility clamp.

EXPERIMENT 9: CALORIMETRY   109

5. Attach the clamp onto a ring stand and lower the test tube into the water being heated (See Figure 9.1). The portion containing the metal should be surrounded by water. You do not have to wait for the water to boil before placing the test tube in the water. Do it as soon as you can so that the temperature of the test tube and the metal can begin equilibrating with that of the hot water.
6. Heat the water to boiling and ***continue*** to heat for at least 10 minutes more.

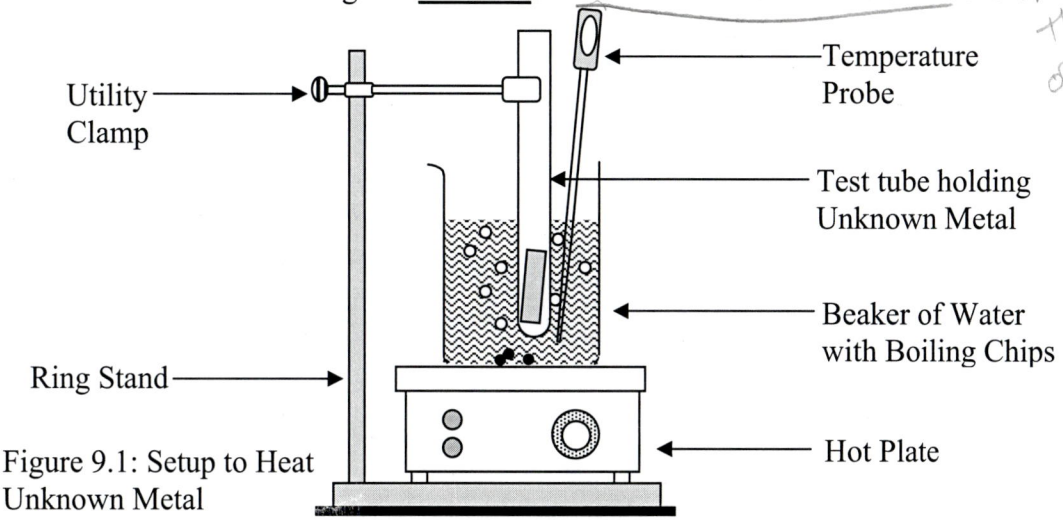

Figure 9.1: Setup to Heat Unknown Metal

(Labels: Utility Clamp, Ring Stand, Temperature Probe, Test tube holding Unknown Metal, Beaker of Water with Boiling Chips, Hot Plate)

7. **Setting up the calorimeter:** Meanwhile, one of the partners should be preparing the calorimeter as follows:
   a. Obtain a calorimeter (double nested Styrofoam cups). Examine both cups to ensure there are no holes on the bottom of the cups. Check to see that they are dry.
   b. Record the mass of the calorimeter (the nested cups).
   c. Measure approximately 75 mL of deionized water and carefully add it to the calorimeter, being careful that the outside of the calorimeter does not get wet.
   d. Record the combined mass of the calorimeter and water, then place it in a 250-mL beaker to keep it safely upright, far away from the hot plate.
   e. Cover the calorimeter with the lid and insert a SECOND temperature probe through one of the holes in the lid. Do not use the same temperature probe as the one being used for the hot water. After a minute, check to see whether the temperature in the calorimeter has stabilized. Meanwhile study Step 9.

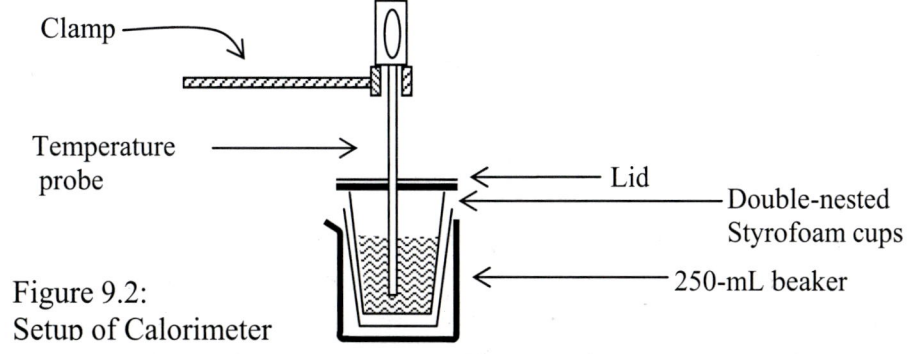

Figure 9.2: Setup of Calorimeter

(Labels: Clamp, Temperature probe, Lid, Double-nested Styrofoam cups, 250-mL beaker)

8. Wait until the metal has been heated for at least 10 minutes, and the temperature of the water in the calorimeter has stabilized. At this point record the temperature of the boiling water in the beaker. We will assume this is the temperature of the hot metal.

110   EXPERIMENT 9: CALORIMETRY

and it is considered the *initial temperature of the metal*: $T_{initial}$ of metal. Next record the temperature of the water in the calorimeter (at room temperature). This is the *initial temperature of the water*, $T_{initial}$ of water.

9. Slide the heated metal into the calorimeter and **quickly** cover with the lid. Gently swirl the contents of the calorimeter and measure the **maximum temperature** (T_{final}) reached by the water. This is the T_{final} for both the metal and the water in the calorimeter.

> TIPS to minimize experimental error:
> - Transfer the metal quickly to minimize loss of heat to the surrounding air, however, do not allow water to splash out.
> - Swirling must be thorough enough that the hot metal transfers its heat uniformly to the water as quickly as possible; however, avoid splashing the water onto the lid.

10. Take the metal out of the calorimeter and dry it thoroughly with paper towels. Dry the calorimeter with paper towels as well.
11. Repeat Steps 3 through 9 and record the data under Trial #2. Be sure to add more water to the beaker on the hot plate to make up for loss due to evaporation, and bring the water back to boiling.
12. When you are finished with both trials, dry your metal thoroughly with paper towels before returning it to your instructor. Check to make sure there are no boiling chips in the sink.
13. Complete calculations on the Calculations & Results pages and draw your conclusions as to the identity of your unknown metal.

Sample Data Table for Part I: Specific Heat of an Unknown Metal

Unknown Metal Code #: ___
Appearance of Metal (shape and color): _____

	Trial #1	Trial #2
Mass of Metal (g)		
Mass of Calorimeter + Water (g)		
Mass of Empty Calorimeter (g)		
Mass of Water in Calorimeter (g)		
Initial Temp of Water (°C)		
Initial Temp of Hot Metal (°C)		
Final Temp of Water & of Metal (°C)		

Sample Calculations for Specific Heat of Unknown Metal: *Your instructor will take you through these calculations in the pre-lab. Take careful notes so that you can do the same calculations for your own set of data after the experiment.*

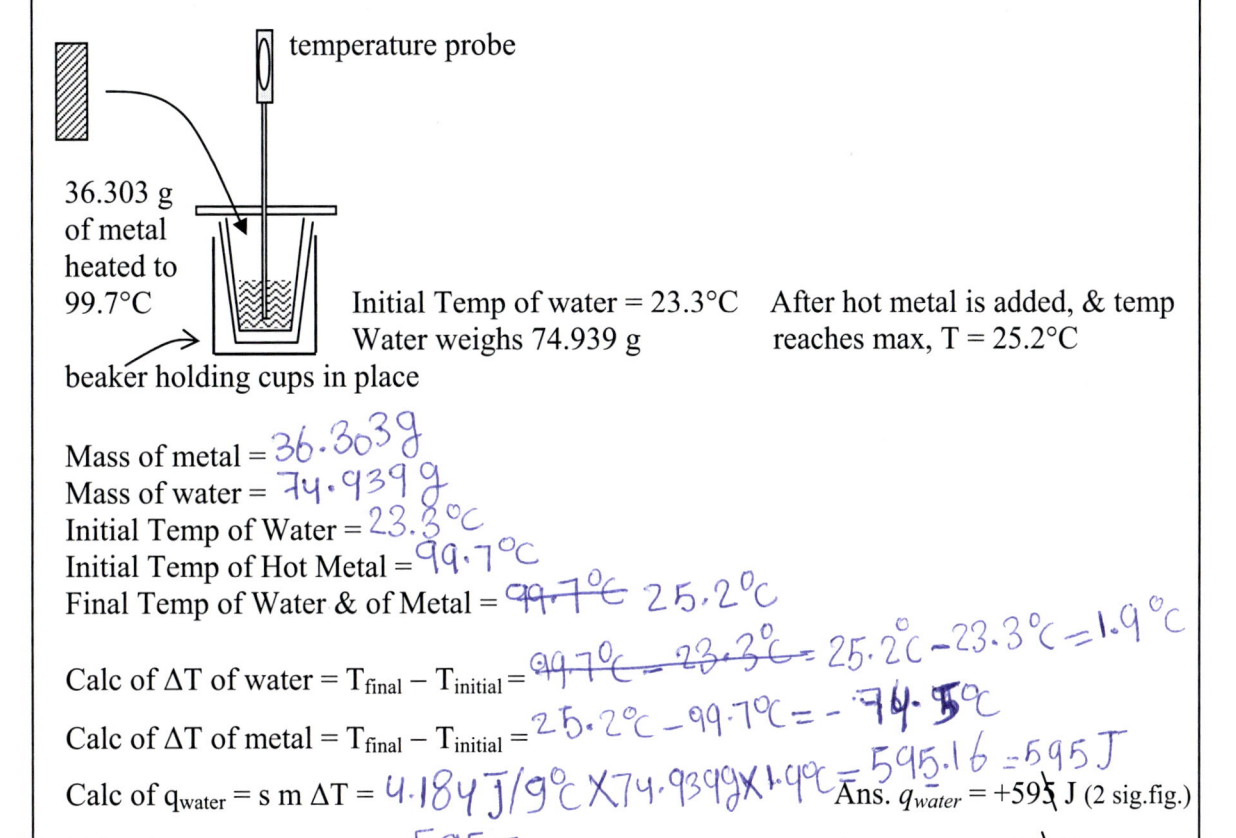

temperature probe

36.303 g of metal heated to 99.7°C

Initial Temp of water = 23.3°C
Water weighs 74.939 g

After hot metal is added, & temp reaches max, T = 25.2°C

beaker holding cups in place

Mass of metal = 36.303 g
Mass of water = 74.939 g
Initial Temp of Water = 23.3°C
Initial Temp of Hot Metal = 99.7°C
Final Temp of Water & of Metal = ~~99.7°C~~ 25.2°C

Calc of ΔT of water = $T_{final} - T_{initial}$ = ~~99.7°C – 23.3°C~~ = 25.2°C – 23.3°C = 1.9°C

Calc of ΔT of metal = $T_{final} - T_{initial}$ = 25.2°C – 99.7°C = –74.5°C

Calc of q_{water} = s m ΔT = 4.184 J/g°C × 74.939 g × 1.9°C = 595.16 = 595 J   Ans. q_{water} = +595 J (2 sig.fig.)

Calc of q_{metal} = – q_{water} = –595 J   Ans. q_{metal} = –595 J

Calc of s_{metal} = $\dfrac{q_{metal}}{m_{metal} \Delta T_{metal}}$ = $\dfrac{-595}{36.303 \times -74.5}$ = 0.22   Ans. s = 0.22 J·g⁻¹·°C⁻¹

Conclusion: What is the metal in this example?

Part II: Molar Heat of Neutralization

1. Take two clean and dry 250-mL beakers and label one "HCl" and the other "NaOH." Similarly, label two clean and dry 50-mL graduated cylinders.
2. Using the labeled 50-mL graduated cylinders, measure out exactly 50.0 mL of the 1.00 M HCl, and exactly 50.0 mL of 1.00 M NaOH solution. If you plan to use droppers to help you measure out the volumes precisely, label these also so you do not mix them up either.
3. Check to ensure the calorimeter, temperature probe and lid are dry.
4. Transfer the 50.0 mL HCl solution from the grad cylinder into the calorimeter, and place the calorimeter in a beaker as shown in Figure 9.2 so that the calorimeter would not topple over. Adjust the clamp so that the temperature probe is not touching the bottom or sides of the calorimeter, but extends below the liquid level.

EXPERIMENT 9: CALORIMETRY

Replace the lid as shown in the figure. Wait 5 minutes for the temperature of the system to stabilize and record it as the "initial temperature." Using a second temperature probe, record the initial temperature of the 50.0 mL of NaOH in the graduated cylinder.

5. Remove the calorimeter lid, quickly pour in the 50.0 mL of the NaOH solution, replace the cover and ***immediately*** begin to swirl the contents of the calorimeter to mix thoroughly. Keep an eye on the temperature and record the maximum temperature as the "final temperature."
6. Thoroughly rinse the inner Styrofoam cup, temperature probes and stirrer with deionized water and gently wipe them dry before repeating steps 2 through 5 for Trial 2. (The grad cylinders and droppers do not have to be dry, but you must be careful you do not get the "HCl" and "NaOH" apparatus mixed up!
7. Complete the calculations on the Calculations & Results page.

CLEANUP: Be sure to rinse the temperature probes with water and wipe them dry. Check to see that your temperature probes are OFF before returning them to the side shelf. Do not discard the Styrofoam cups and lid. Return them to the side shelf after cleaning them.

Sample Data Table for Part II: Molar Heat of Neutralization

	Trial #1	Trial #2
$T_{initial}$ of HCl solution (°C)		
$T_{initial}$ NaOH solution (°C)		
Average $T_{initial}$ (°C)		
Total Volume After Mixing (mL) (Vol HCl soln + Vol NaOH soln)		
T_{final} (Maximum T After Mixing) (°C)		
ΔT (T_{final} − Average $T_{initial}$) (°C)		

EXPERIMENT 9: CALORIMETRY

Sample Calculations for Part II: Molar Heat of Neutralization of HCl and NaOH
NOTE: Numbers used here are fictitious and the ΔH values are nowhere near the correct volume. The # sig. fig. is consistent with what is presented HERE and not necessarily what students will be actually dealing with. Do not blindly use the same # sig. fig. as shown here.

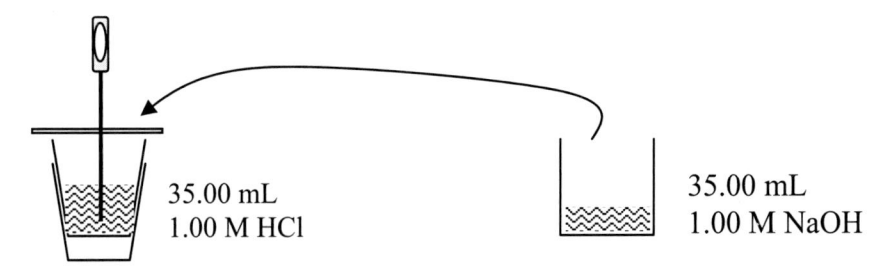

35.00 mL 1.00 M HCl   35.00 mL 1.00 M NaOH

$T_{initial}$ = 25.2°C ⟶ T_{final} = 29.0°C

As your instructor goes through the calculations in pre-lab it would be wise to take careful notes on your own paper. Do not just scribble them on this page.

The heat released in the neutralization (q_{neutr}) will heat up the total solution inside the calorimeter. What is the mass of the solution (m_{soln}) in the calorimeter?

$$HCl(aq) + NaOH(aq) \longrightarrow NaCl(aq) + H_2O\ (l)$$

The neutralization reaction is very fast. Almost immediately the only substances in the calorimeter consist of only NaCl dissolved in water. At this concentration we can assume that the density of this solution is the same as that of water, 0.997 g/mL, and the specific heat of the solution is also the same as that of water, 4.18 J·g^{-1}·°C^{-1}.

Since there are multiple steps in the calculations, always keep at least one extra significant figures until you reach the final answer before rounding off properly.

Total Volume of Solution After Mixing =
Mass of Solution After Mixing (from total volume and density) =
$T_{initial}$ of Reactants =
T_{final} (of Soln) =
Calc of $\Delta T_{soln} = T_{final} - T_{initial}$ =
$q_{soln} = s_{soln}\ m_{soln}\ \Delta T_{soln}$ =
Convert q_{soln} to kJ =   Ans. + 1.10 kJ
$q_{neutr} = -q_{soln}$ =   Ans. –1.10 kJ

To calculate *molar heat of neutralization*, we need the # mol limiting reactant.
In this case, either HCl or NaOH can be used as the limiting reactant.
How do we determine # mol HCl? What information do we have concerning the HCl?
We know its volume and its molarity.
Volume of HCl used in neutralization =
Molarity of HCl used in neutralization =
mol HCl =   Ans. 0.0350 mol HCl

$$\text{molar heat of neutralization} = \frac{\text{heat transferred in each trial}}{\text{\# mol HCl in each trial}} = \frac{q_{neutr}}{\text{\# mol HCl}} =$$

Ans. Molar heat of neutralization =   Ans. –31 kJ/mol HCl

EXPERIMENT 9: CALORIMETRY

Pre-Lab Exercise:
1. In Part I of the experiment, which do you expect to be larger, $T_{initial}$ or T_{final} of the water? Based on your answer, do you expect ΔT_{water} to be positive or negative? Explain.
2. In Part I of the experiment, would you expect q_{metal} to be positive or negative? Would you expect q_{water} to be positive or negative? Explain.
3. Consider the two parts of the experiment. What is producing the heat measured in Part I? What is producing the heat measured in Part II? Explain your answers.
4. How are the terms "heat of neutralization" and "molar heat of neutralization" different? Include in your answer an explanation why one is an extensive property and the other is an intensive property.
5. In this experiment q equals ΔH? What are the experimental conditions that allow us to equate them?
6. In Part I of the experiment, we utilize the equation $q_{metal} = s \, m \, \Delta T$. If we change the mass of the metal, what would you expect to change (q? s? m? ΔT?). Explain your answer.

Post-lab Questions:
1. Examine the initial and final temperatures in Part I. Explain how the temperatures tell you what type of reaction was involved (endothermic or exothermic). Are the signs of your q_{water} and q_{metal} consistent with this? Explain.
2. In Part I, we see that copper and zinc have the same specific heat (See table in the Introduction.) If you obtained an experimental value of 0.39 J·g^{-1}·°C^{-1}, how might you determine which metal you have as an unknown? Explain.
3. We assumed that no heat is lost to the surroundings beyond the nested coffee cups. In Part I, obviously there would have been some loss in heat as the hot metal is transferred to the calorimeter. How does that unavoidable heat loss affect your calculated specific heat of the metal? Would your calculated specific heat be too high or too low due this error? Explain fully.
4. In Part II, we assume that the density and specific heat of the solution is the same as that of water. What justifications do we have to make that assumption? Explain.
5. Using the molar heat of neutralization **obtained in your experiment** (assuming it is correct), calculate how much heat you would expect to be produced if you mixed 50.0 mL of 0.250 M HCl with by 50.0 mL of 0.250 M NaOH. Show your calculations. (Hint: How many moles of HCl are involved?)

Experiment 10: ATOMIC SPECTROSCOPY

Purpose:
Part I: Verify that Bohr's theory on the structure of the H atom is correct, and determine the electron transition corresponding to each spectral line in the H emission spectrum.
Part II: Identify metal ions in two unknown aqueous solutions based on a flame test.

Performance Goals:
- Observe emission spectrum of hydrogen
- Prepare a calibration plot based on the helium emission spectrum
- Predict wavelengths of spectral lines in the H emission spectrum based on Bohr's theory.
- Assign the spectral lines in the H emission spectrum
- Use a flame test to identify the cations in two unknown salts

Introduction:

Each element that makes up the natural world has its own unique signature in the form of the *electromagnetic* (EM) radiation that it absorbs or emits. This signature is referred to as the absorption or emission spectrum of the element. So unique is this signature that astronomers can determine the chemical composition of distant stars by analyzing the light they emit.

Early attempts to construct a theory that would explain and predict spectral properties began by considering the element hydrogen. Atoms of hydrogen are the simplest atoms that exist in nature consisting of only one proton and one electron and it seemed prudent to deal first with simple systems. In 1900 a German physicist by the name of Max Planck published a revolutionary paper showing that the energy associated with EM radiation could be determined from the frequency or wavelength of the radiation using the following equation.

$$E = h\nu \qquad \text{Equation 1}$$

where E = energy of EM radiation
h = Planck's constant = 6.626×10^{-34} J·s
ν = frequency of EM radiation in units of s^{-1} (cycles per second) or Hz (Hertz)

The frequency of the radiation is inversely proportional to the wavelength and can be given by the following equation.

$$\nu = \frac{c}{\lambda}$$

where c = speed of light = 2.998×10^8 m·s^{-1}
λ = wavelength of EM radiation in units of m

Substituting for the frequency, ν, in Equation 1 gives the following equation.

$$E = \frac{hc}{\lambda} \qquad \text{Equation 2}$$

Planck's discovery set the stage for the first successful attempt to explain and predict the spectral properties of hydrogen. This was accomplished by Niels Bohr in 1913. Bohr pictured the hydrogen atom as a miniature solar system in which the tiny electron orbits about the much larger proton. According to Bohr, electrons were restricted to occupying only certain allowed orbits, each of which was denoted by a whole number ranging from 1 to ∞ (infinity). These numbers were called principal quantum numbers. Bohr showed how the energy of an electron in a particular orbit could be calculated from the principal quantum number (n) of the orbit.

$$E_n = \frac{-2.180 \times 10^{-18} \text{ J}}{n^2}$$
Equation 3

where n = principal quantum number (ranging 1 to ∞)
E_n = energy of electron for a particular quantum number, n

The electron orbit where n = 1 is the smallest orbit and has the least energy. When the single electron of hydrogen is in this orbit, the atom is said to be in its *ground state*, being in its most stable state. This electron can jump to orbits of higher energy where n = 2, 3, 4 or higher if it absorbs sufficient energy from some external source. Electrons are then said to be in an *excited state*. Electrons in excited states tend to return to the ground state and emit energy in the form of EM radiation when they do. The amount of energy emitted or gained by an electron (ΔE) when it undergoes a jump from one allowed energy state to another can be calculated by first determining the energy of each state using Equation 3 and then finding the difference in energy (ΔE) between the two energy states. Since emission is always due to the transition from a higher state to a lower state, ΔE for such transitions is always E of the higher state minus E of the lower state, and it will always be a positive number.

$$\Delta E = E_{high} - E_{low}$$
Equation 4

This amount of energy (ΔE) released is equal to the energy (E) of a photon of the light emitted (that is, ΔE = E_{photon}). Max Planck had previously established an equation (Equation 2) showing the relationship between the energy (E) of a photon and the wavelength (λ), and frequency (ν) of the light. Putting Equations 2 and 4 together Bohr was able to predict the wavelength and frequency of the light emitted for each transition of the electron of hydrogen as it jumps from one level to another.

$$\Delta E_{electron} = E_{photon} = h\nu = \frac{hc}{\lambda}$$

Note: Usually the subscripts are dropped and the equation becomes simply

$$\Delta E = E = h\nu = \frac{hc}{\lambda}$$

This equation, when rearranged to solve for ν or λ, becomes

$$\nu = \frac{\Delta E}{h}$$

or $$\lambda = \frac{hc}{\Delta E}$$
Equation 5

As indicated by Equation 5, the wavelength (λ) of EM radiation is inversely proportional to its energy. Short waves have high energy and long waves have low energy. The human eye is only capable of seeing a narrow region (ranging in wavelength from 400 nm to 750 nm) of EM radiation called visible light. Visible light is made up of light of different colors but the human eye sees the light as white because it cannot separate the colors. If visible light is passed through a prism or diffraction grating it produces a continuous band of colors (red, orange, yellow, green, blue, indigo, and violet). But, when atoms are excited (in a gas discharge tube or in a flame), and the light emitted is passed through a prism or diffraction grating, it does not produce a continuous band of colors. Instead, a spectrum with discrete lines called a *line spectrum* or *emission line spectrum* is produced. Each element has its own unique set of EM radiation absorption and emission pattern, because the energy difference between one energy level to another (ΔE) is different for each element, giving rise to lines with different colors being produced. For example, when sodium compounds are placed in a flame they produce a bright yellow flame while strontium compounds give brilliant reds. This is how we get the spectacular display of colors in fireworks. By examining the emission pattern from the stars we can determine what elements are there without traveling to the stars to obtain a sample.

In Part I of this experiment you will measure the wavelengths of the light emitted by hydrogen in the visible region of the EM spectrum and compare them with those calculated from Bohr's theory. If the percent error is small we can conclude that Bohr's theory is supported by experimental evidence.

Unfortunately, the observed wavelengths using our spectroscopes may not be accurate, but they can be corrected by first preparing a calibration curve using the known and observed wavelengths of the helium atom as a guide. The calibration curve is done by recording the observed wavelengths of the spectral lines of helium and plotting the known helium wavelengths (y-axis) against the observed helium wavelengths (x-axis). Each spectroscope will produce its own calibration curve. It is therefore important that you use the same spectroscope for both the hydrogen and helium emission spectra.

Once you have prepared the calibration curve, you will correct the observed values for hydrogen by reading them off your calibration curve. By comparing each corrected observed wavelengths of hydrogen with those predicted by Bohr's theory, you will then identify each spectral line with the transition that produced it and verify whether Bohr's theory is indeed valid.

In Part II, two unknown metal ions will be identified by comparing the color of light they emit with those of selected known metals ions using a flame test.

Equipment/Materials:

Spectroscope, electrodes receptacle, helium gas discharge tube, hydrogen gas discharge tube, known and unknown salt solutions, de-ionized water, gas burner, gas striker, a ceramic spotting plate, a piece of copper wire attached to a cork, wire cutter, and a 100- or 150-mL beaker

120   EXPERIMENT 10: ATOMIC SPECTROSCOPY

Procedure:
Part I: Spectral Lines of Helium & Hydrogen
A. Preparation of a Calibration Curve for the Spectroscope Using the Helium Spectrum

First check to make sure the power is off, then place a helium gas discharge tube in the receptacles of the light source. Turn on the light source. The helium tube should now be illuminated. Adjust the height of the spectroscope or the light source (with wooden boards or whatever is provided to be used) so that the spectroscope is positioned at the middle of the helium tube where the light is brightest. While looking through the eyepiece, aim the spectroscope at the light source by slowly moving it back and forth horizontally until a bright beam of light enters the slit at the front of the unit. The position should be fine-tuned so that this light beam is aimed straight at your eye. To the right of the slit, you will observe a virtual image of the helium spectrum displayed on a calibrated scale.

The scale you will see is shown below. Take a moment to first examine the scale, which is in nanometers (nm) (same as millimicrons, mμ), and determine ***how many sig. fig.*** can be read from this scale. Record in your data sheet the wavelength reading, color for each bright band of light observed. It will make the analysis easier later on if you record the wavelengths in numerical ***decreasing*** order in "Table 10.1: Helium Emission Data." That means start with the line on the far right and work towards the left. Turn off the light source when you are finished.

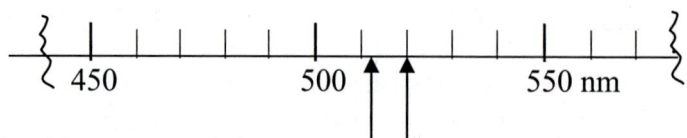

How should you record the positions of the two arrows in this figure? How many sig.fig. should they have? Check your answer with your instructor before proceeding.

You will see more lines than are listed in Table 10.1. Pick the brightest lines with colors that match. You ***must*** have a wavelength for each color line in the table.

B. Observation of Hydrogen Spectrum

With the power off, remove the helium tube from the light source and substitute a hydrogen tube.

> **CAUTION! THE HELIUM TUBE WILL BE HOT! USE FOLDED PAPER TOWELS TO HANDLE IT.**

Turn on the light source and follow the same procedure as in Part A in observing and recording the hydrogen spectrum. (Remember that it is important that the same spectroscope be used for all wavelength measurements on both H_2 and He.) You should be able to see at least three lines. If you see a fourth line, record its color and

EXPERIMENT 10: ATOMIC SPECTROSCOPY   121

wavelength as well. Again, you will find it wise to do this systematically in numerical decreasing wavelengths in "Table 10.2: Hydrogen Emission Data."

Analysis of Data in Part I:

A. Analysis of the Helium Spectral Lines
The accepted wavelength values and colors for helium are listed in Table 10.4. Next to each accepted value write in the observed wavelength for the value that most closely matches it. Make sure the color matches also. BEFORE YOU BEGIN PREPARING YOUR GRAPH, TABLE 10.4 MUST BE COMPLETED!

Preparation of the Calibration Graph Using Excel 2010
The objective is to prepare a graph using helium spectral lines to convert observed wavelengths to corrected wavelengths. The instructions below utilize Excel 2010, which is the version used on all CCBC college computers. If laptops are provided, you are expected to complete this graph in the lab. The instructions are brief as you have already been introduced to the use of Excel 2010 in an earlier experiment.

*We begin with formatting the first two cells in **Row 1** so that all content of each cell is visible within the cell by displaying it to multiple lines (**Wrap text**). The objective is to minimize the width of the data table so that there is more space for the graph in the final printout.*

1. Click on the **Home** tab.
2. Highlight Row 1, Cells A and B. ——— Then click on **Wrap text**.

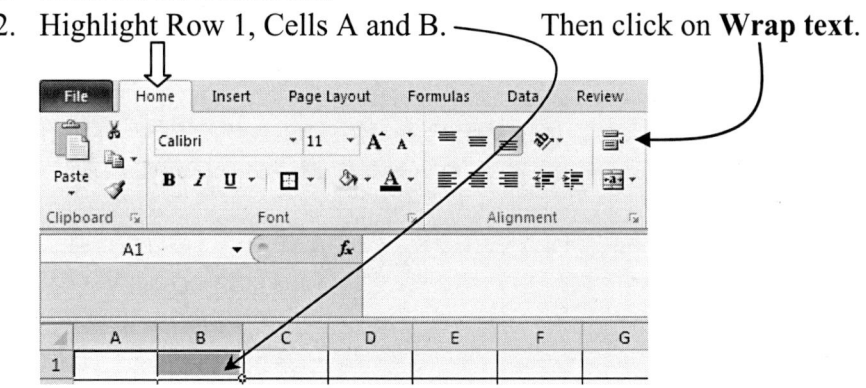

3. In Row 1, label Column A ***Observed Wavelength (nm)*** and label Column B ***Corrected Wavelength (nm)***.
4. Enter in Column A, the Observed Wavelengths of Helium (from Table 10.4).
   Enter in Column B, the corresponding Accepted Wavelengths of Helium (from Table 10.4).
5. After all the data has been entered, highlight both columns, including Row 1, down to the last row that contains an entry.
6. Click on the **Insert** tab, on **Scatter**, and then on **Scatter with only Markers**.

122   EXPERIMENT 10: ATOMIC SPECTROSCOPY

[handwritten: Y = 1.0854x − 28.533]

> Before proceeding, examine your graph. If there are any "outliers" first check to make sure it is not due to an error in entering the data. If it is due to some other error, it is most likely you matched your observed helium wavelengths with the wrong accepted helium wavelengths. Get help from your instructor. You cannot proceed until this is resolved.

7. Place your cursor anywhere on the spreadsheet (not on the graph) and click on **Page Layout** tab, **Orientation** and select **Landscape**. The dotted lines that appear indicate the size of the page you will be printing. Click on the frame of the graph and then resize the graph so that it is as large as possible without letting it spill beyond the dotted lines.
8. Add the line of best fit by clicking **Layout, Trendline,** and selecting **Linear Trendline**.
9. To obtain the equation for the trendline, click again on **Layout, Trendline,** and select **More Trendline Options** at the very bottom. The **Format Trendline** window will then appear.
10. Place a check mark at **Display Equation on chart,** and click on **Close.**
11. If necessary, move the equation to a position where it can be read easily (such as just below the title). This is done by clicking on the equation once and then dragging it to the desired position.
12. Next remove the legend (as it does not apply to this graph) by clicking on **Legend** and selecting **None.**
13. Change the title by first clicking (***only once***) on the default title and then type in your own title: Calibration Graph for Spectroscope: Corrected vs. Observed Wavelengths. When you press **ENTER,** your title will appear on the graph. Alternatively, you can highlight the default title and type in your own title.
14. Next click on **Axis Titles**, then select **Primary Horizontal Axis Title,** and **Title Below Axis**. Type in a title for the x-axis: Observed Wavelengths (nm). Press **ENTER.**
15. Click on **Axis Titles,** and then select **Primary Vertical Axis Title,** and **Rotated Title**. Type in a title for the y-axis: Corrected Wavelengths (nm) Press **ENTER.**
    As you have probably noticed, the data are bunched up in a small area of the graph. We will now adjust the minimum and maximum for the x- and y-scales.
16. Click on **Axes,** select **Primary Horizontal Axis** and then on **More Primary Horizontal Axis Options**. This is where you will adjust for the x-scale (Observed Wavelengths). Select **Fixed** to allow you to change the default settings.
17. Remember the minimum must be smaller than your smallest x-value, and the maximum must be larger than your largest x-value. Select **Fixed** to allow you to change the default settings. If your lowest x-value is 440, and the maximum, around 660, try setting the minimum = **400** and maximum = **700**. This may not work for your data. You must choose settings that fit your data. Adjust the Major Unit and Minor Unit as needed. Do not forget to go down to **Minor tick mark type** and change "**None**" to "**Outside.**" When you are finished, click on **Close.**
18. Click on **Axes,** select **Primary Vertical Axis** and then on **More Primary Vertical Axis Options**. Remember to select **Fixed** to change the default settings. Enter your

EXPERIMENT 10: ATOMIC SPECTROSCOPY   123

Minimum, maximum, Major unit, Minor unit. Go down to **Minor tick mark type** and change "**None**" to "**Outside.**" When you are finished, click on **Close**.

19. Click on **Gridlines**, select **Primary Horizontal Gridlines** and **Minor Gridlines**.
20. Click on **Gridlines**, select **Primary Vertical Gridlines** and **Minor Gridlines**.
21. In the area above the graph (such as in Cells F1 and F2), enter your course number, course section, semester, and names of students who prepared the graph. For example: *CHEM 122 - Sec CM2 - Spring 2006   Jane Smith and John Doe* (It is important to include your names so that somebody else does not pick up your printout!) Drag and/or resize the graph if necessary in order to make space for the above information.
22. Highlight the entire page. Click on **Page Layout** tab, **Print Area**, and select **Set Print Area**.
23. Click on **File, Print**, and then **Print Preview** to double check that everything fits on the page, then click on **Print**. (Print 2 copies, one for each partner.)
24. Both partners must copy the equation of the trendline into his/her lab notebook.

B. Analysis of the Hydrogen Spectral Lines

1. **Determining Correct Wavelengths for the Hydrogen Spectral Lines**
   The Observed H Wavelengths in Table 10.2 are x-values. From the equation of the trendline, calculate y for each of the Observed H wavelengths and record the answers into Table 10.2 as Corrected H Wavelengths. These values should have the same significant figures as your observed wavelengths. You cannot have more precision in calculated values than in observed values. (Hint: These Corrected H Wavelengths should be in the general ballpark as the Observed H Wavelengths. In other words, if your Observed value is 650 nm, the Corrected value cannot be 1870 nm. If they differ significantly, check your calculations!) Show your calculation setups on a separate piece of paper to be submitted.

2. **Matching Each Spectral Line of Hydrogen with Its Electron Transition**
   The mechanism by which atoms produce light involves electrons undergoing transitions from higher quantum levels to lower ones. A transition from n = 7 to n = 2 should be entered as n = 7→2, with the arrow showing the direction of the transition.

   Copy the "Corrected Observed H Wavelength" from Table 10.2 into Table 10.5. Next, match each of these corrected hydrogen wavelengths to the calculated hydrogen wavelength (from Appendix 4) that is closest in value to it. Record in the next column, the matching calculated hydrogen wavelengths. In the next column, write down the corresponding transition (e.g. n = 5→2, or n = 4→3 etc.).

124   EXPERIMENT 10: ATOMIC SPECTROSCOPY

3. **Analysis of the Accuracy of the Observed H Lines.**
   Calculate the error and % error of each of the "Corrected Observed H Wavelengths" using the theoretical wavelength provided in Appendix 4 as the accepted value. Refer to Experiment 2 on how to calculate error and % error. Enter the values in Table 10.5. What information does the % error give you?

Part II: Identification of Unknowns Using a Flame Test

A. Flame test for metal cations

> **WEAR YOUR SAFETY GOGGLES AT ALL TIMES!**

1. Your instructor will assign you two unknown samples in small vials, each containing one unknown metal cation in aqueous solution. Record the code number of the unknowns onto your Data Sheet.

2. Clean and dry a spot plate thoroughly then place it on a horizontal surface near where the gas burner will be used and add 4 to 6 drops of each of the following solutions to separate cavities on the plate: 1M $NaCl(aq)$, 1M $KCl(aq)$, 1M $BaCl_2(aq)$, 1M $Ca(NO_3)_2(aq)$, 1M $LiCl(aq)$, 1M $CuCl_2(aq)$, 1M $NiCl_2(aq)$, 1M $SrCl_2(aq)$. Make a diagram on a separate sheet of paper showing which solution is contained in each cavity.

3. Clean the copper wire thoroughly under the tap water, and then rinse it with deionized water. Fill a large beaker with deionized water. This water will be used to rinse the copper wire between samples. Tap water cannot be used as it contains sodium ions which burn with a bright yellow-orange color and will mask any other colors that may be emitted by your test samples.

4. Light the gas burner and adjust the flame to give a hot intense blue flame about 12 cm high (about 5 inches) with an inner blue cone approximately 5 to 6 cm high (about 2 inches high). **If you have long hair, remember to tie your hair back!**

5. Begin by testing your copper wire. Hold it by the cork as the wire will get very hot. Place the tip (only) of the wire into the hottest part of the flame for about 3 or 4 seconds. Note the color of the flame as the wire glows red hot. The flame should show only a very pale yellow color. If it gives a green or crimson red color flame, it is contaminated. Hold it longer in the flame until the color is gone. If the color persists, ask your instructor to cut off the tip of the wire and test again.

6. When you are sure that the copper wire is clean, dip the tip in one of the known solutions, and again place the tip in the hottest part of the flame for about 3 or 4 seconds. Record the color of the flame, and whether it is fleeting or persistent.

Especially when two cations have similar colors, you will have to distinguish between them by noting whether the color disappears quickly (said to be "fleeting") or lasts a relatively long time (said to be "persistent").

7. Before you test your next sample, the wire must be cleaned so that it does not contaminate your sample. Heat the tip of the wire until the previous sample has been burned off completely (until the flame no longer shows the color of the previous sample). Allow the wire to cool down for about a minute, and then swish it around in the beaker of deionized water. Heat the tip of the wire again to ascertain that the wire is indeed clean (burning with a flame with no other color than the pale yellow). IF YOU ARE UNABLE TO GET RID OF THE PREVIOUS SAMPLE IN THIS MANNER CONSULT WITH YOUR INSTRUCTOR.

Test $SrCl_2$ last as it tends to stick to the wire and is hard to clean off. Before you go on to your unknowns, have your instructor snip off the end of the wire so that you can start with a fresh section of the wire.

Identify the unknown metal ions by comparing the flame color with those of the known metal ions. Remember to clean the wire between each flame tests as you had done for the known solutions. You will be graded on your accuracy of identification, so repeat this procedure until you are satisfied with your conclusion. (HINT: DO YOU HAVE ANY EVIDENCE FOR THE IDENTITY OF THE ANION?)

8. Write into Table 10.3, the formula of the metal ion that you conclude is present in each of your unknowns.

Prepare your Data Tables in your lab notebook.

PART I.

Table 10.1: Helium Emission Data

Observed Color	Observed Wavelength (nm)
red	
yellow	
green	
green	
blue-green	
blue-violet	

Table 10.2: Hydrogen Emission Data

Observed Color of H Line	Observed H Wavelength (nm)	Corrected H Wavelength (from the Excel Trendline Equation)

Part II: Identification of Unknowns Using a Flame Test
Remember to record the code numbers of your unknown!

Table 10.3: Observations & Conclusions from Flame Test

SOLUTIONS	METAL ION PRESENT	FLAME COLOR
cleaned copper wire	none	
1M NaCl	Na^+	
1M KCl	K^+	
1M SrCl$_2$	Sr^{2+}	
1M BaCl$_2$	Ba^{2+}	
1M Ca(NO$_3$)$_2$	Ca^{2+}	
1M LiCl	Li^+	
1M CuCl$_2$	Cu^{2+}	
1M NiCl$_2$	Ni^{2+}	
Code # of unknown = _____	Conclusion:	Observation:
Code # of unknown = _____	Conclusion:	Observation:

Pre-Lab Exercise:

Your instructor will indicate whether you are to hand in answers to these questions or have you take a pre-lab quiz with similar questions.

1. Calculate the energy of an electron of the hydrogen atom when it is in the energy levels indicated:
   a) n = 3   b) n = 6   c) n = ∞
2. Using Bohr's Equation calculate the energy change experienced by an electron when it undergoes transitions between the following energy levels.
   a) n = 6 → 3   b) n = 6 → ∞
3. Using Bohr's Theory (**not** Rydberg's equation) calculate the wavelength, in units of nanometers, of the electromagnetic radiation emitted for the electron transition n = 6 → 3. (1 m = 10^9 nm).
4. In what region of the electromagnetic spectrum (UV? microwave? etc.) would the electromagnetic radiation occur for the transition n = 6 → n = 3? (Hint: See the section on Electromagnetic Radiation in your chemistry text.)
5. In the flame test, why must Sr^{2+} be tested last?
6. In the flame test if two ions have similar colors what will help you distinguish one from the other?

ELECTROMAGNETIC RADIATION SPECTRUM
Wavelength in nanometers

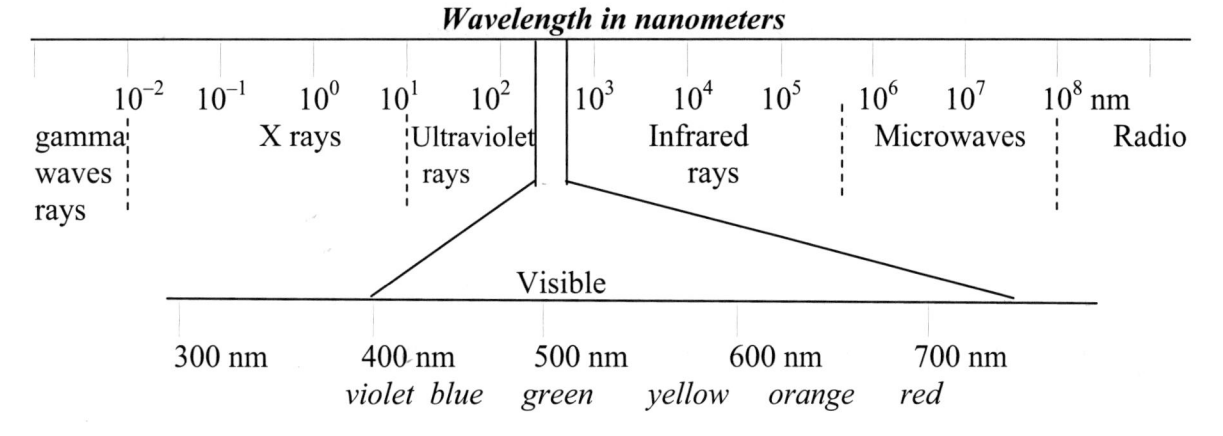

```
   10⁻²  10⁻¹  10⁰  10¹  10²  | 10³  10⁴  10⁵  10⁶  10⁷  10⁸ nm
  gamma       X rays   Ultraviolet   Infrared     Microwaves    Radio
  waves                   rays         rays
  rays
                              Visible
         300 nm      400 nm    500 nm     600 nm      700 nm
                  violet blue  green   yellow orange  red
```

Post-Lab Questions: (Please <u>type</u> your answers.)

1. What was the goal in Part I of the experiment? Write a conclusion to this part of the experiment, showing how you have reached the goal specified. Be as specific as you can.

2. In your own words explain fully, <u>as precisely as you can</u>, why hydrogen is emitting a blue-green light in this experiment.

3. Consider how your observations would be different if your spectroscope were accurate:
   a) How would you expect the <u>observed</u> wavelengths of helium to be different from the <u>observed</u> wavelengths you have recorded in Table 10.4? Be specific in your answer.
   b) What would you expect the value of the slope of your calibration graph to be?
   c) What would you expect the value of the y-intercept of your calibration graph to be?
   d) What would the trendline equation be?

4. Look up in the Internet what is known as the Lyman Series.
   a) Which transitions do the lines in the Lyman Series correspond to?
   b) Would you expect the emission lines in the Lyman Series to be in the visible region? Explain fully how you came to your conclusion. Include the reference source from the Internet in the proper format, including the access date.

5. Explain what a flame test has to do with spectroscopy.

6. (For this question only, answers are to be handwritten rather than typed.) Consider the electron of the H atom that undergoes the transition of n = 9 → 6.
   a) How much energy is released? Show your calculations carefully. As usual, all setups must include units.
   b) What would be the wavelength (in nm) of the light emitted? Show your calculations carefully. As usual, all setups must include units.

Experiment 11: MOLECULAR GEOMETRY & POLARITY

Purpose: Determine how the number of electron groups about a central atom affects the shape of a molecule and how the shape affects molecular polarity

Performance Goals:
- Construct Lewis structures
- Use the VSEPR method to determine 3-D shapes
- Predict hybridization of central atoms in simple molecules
- Use molecular models to construct 3-D structures from Lewis structures
- Determine molecular polarity

Introduction:

Molecular Geometry

Molecular geometry refers to the 3-D shapes of molecules and polyatomic ions. The shape of a simple molecule or a polyatomic ion with one central atom can easily be predicted from Lewis structures by applying the *valence shell electron pair repulsion* (VSEPR) theory. According to the VSEPR theory, groups of electrons about a central atom are arranged so that repulsion between the groups is at a minimum. A group of electrons could be a single bond, a double bond, a triple bond, a lone pair, or a single electron. A double bond has two pairs of electrons but counted as one group because both pairs are attached to the same atom. Similarly the three pairs of electrons in a triple bond are counted only as one group. It is important to note that the actual shape of a molecule or polyatomic ion is the same as the arrangement of the groups of electrons **only** when all groups of electrons about the central atom are bonding electrons. If one or more groups of electrons are nonbonding (lone pairs, or a single electron), the shape of the arrangement of electrons about the central atom and the actual shape of the molecule are not the same.

For instance, four groups of electrons about a central atom will be arranged in a tetrahedral manner to minimize repulsion, but if two of these groups are bonding electrons and two are nonbonding, the actual shape of the molecule is not tetrahedral but bent. Such is the case for water. According to the Lewis structure there are four groups of electrons (sometimes referred to as *domain*) about the oxygen – two bonding groups and two nonbonding groups. Because only two of the groups are bonding groups, the molecular geometry of a water molecule is described as bent rather than tetrahedral.

Fig. 11.1

      Lewis structure   Tetrahedral arrangement   Molecular geometry
                                of four electron groups   described as "bent"

Note: For geometry purposes both double and triple bonds are considered to have one electron group between the atoms forming the double or triple bond. For example, there are two electron groups around carbon in carbon dioxide (O = C = O), not four. Similarly, there are two electron groups around carbon in hydrogen cyanide (H – C ≡ N:). It has been found that only one of the electron pairs in a double or triple bond influences the geometry.

Another theory, the valence bond (VB) theory approaches molecular geometry from a different angle. According to VB theory, bonds between atoms are formed from overlap of orbitals in the valence shell. The geometry is based on the spatial arrangement of the orbitals. In many instances however, the overlapping of pure s, p, and d orbitals cannot account for the shape and the only way to do it is to assume that the overlapping orbitals on the central atom are *hybrids*. Hybrid orbitals are formed from the mixing of pure s, p, and d orbitals to generate new orbitals called sp, sp^2, sp^3, sp^3d, etc. The spatial arrangement of these orbitals are shown in Table 11.1. The particular hybrid orbital that is formed depends on the proportion of pure (s, p, d, etc.) orbitals in the mix. If one part s and one p are mixed, the hybrid orbitals are called sp, while one part s and two parts p would give rise to sp^2 orbitals, and so on. The number of hybrid orbitals formed depends on the number of pure orbitals that are mixed. For instance, if an s orbital and p orbital are mixed, two sp hybrid orbitals are formed; and if an s and two p orbitals are mixed, three sp^2 hybrid orbitals are formed, and so on. If two hybrid (sp) orbitals are formed, they will be arranged 180° apart; three (sp^2) hybrid orbitals will be 120° apart; four (sp^3) hybrid orbitals will be 109.5° apart, and so on to minimize the repulsion between the orbitals. It is important to note that the spatial arrangement of the hybrid orbitals about a central atom is not necessarily the same as the shape of a molecule or polyatomic ion. Molecular geometry describes the spatial arrangement of the <u>atoms</u> and not of the orbitals: orbitals of water is arranged as a tetrahedron, but the molecular geometry of water is described as bent.

The hybridization of a central atom can be determined from Lewis structures. In a Lewis structure, if there are say two groups of electrons about a central atom, it means that two hybrid orbitals would be required to hold them. Whenever two hybrid orbitals are formed, the hybridization on the central atom is always sp. Similarly, if there are say four groups of electrons about a central atom, it means that four hybrid orbitals would be required to hold them. And, whenever there are four hybrid orbitals, the hybridization on the central atom is sp^3. For example, because there are four groups of electrons on oxygen in the Lewis structure for water, four hybrid orbitals would be required to hold them; and so, the hybridization of the oxygen is sp^3. Table 11.1 gives the shapes of simple molecules and polyatomic ions that are expected from VSEPR and corresponding hybridization of the central atom.

Some shapes such as linear and trigonal planar can easily be represented on a 2-D surface such as on paper or a blackboard. For other shapes such as trigonal pyramidal and tetrahedral, in which the atoms of the molecules are not all in the same plane, special designations to show all bonds must be used. Bonds in the plane of the paper are represented with a regular line (———), bonds projecting forward out of the plane are shown with a solid wedge (◤), and bonds projecting backwards behind the plane are shown with a broken wedge (┈┈). The following examples show how the 3-D structures of linear, trigonal planar and tetrahedral molecules can be represented on paper.

Fig. 11.2

	O=C=O	F–B(–F)(–F)	Cl–C(–Cl)(–Cl)(–Cl)
	linear	trigonal planar	tetrahedral
Hybridization of Central Atom	sp	sp^2	sp^3

EXPERIMENT 11: MOLECULAR GEOMETRY & POLARITY   133

Table 11.1: Relating Groups of Electrons about the Central Atom to Molecular Geometry

# Groups of Bonding electrons	# Groups of Non-Bonding electrons	Total # Groups of Electrons	Hybridization	3-Dimensional Sketch	Molecular Geometry & approx. bond angle (Actual angles differ due to lone pairs requiring more space.)	Example
2	0	2	sp		linear bond angle = 180°	H–C≡N:
1	1	2	sp		linear bond angle = N/A*	⁻:C≡N:
3	0	3	sp²		trigonal planar bond angle = 120°	[:O=C–O:]²⁻
2	1	3	sp²		bent (or angular) bond angle ≈ 120°	[:O=N–O:]⁻
1	2	3	sp²		linear bond angle = N/A*	O=O
4	0	4	sp³		tetrahedral bond angle = 109.5°	H–C(H)(H)–H
3	1	4	sp³		trigonal pyramidal bond angle ≈ 109.5°	H–N(H)–H
2	2	4	sp³		bent or angular bond angle ≈ 109.5°	H–O–H
1	3	4	sp³		linear bond angle = N/A*	[H–O:]⁻

*A bond angle is the angle between two bonds. A species must contain at least three atoms in order to form a bond angle. In this table it is marked as N/A for "not applicable."

134   EXPERIMENT 11: MOLECULAR GEOMETRY & POLARITY

# Groups of Bonding Electrons	# Groups of Non-Bonding Electrons	# Total Groups of Electrons	Hybridization	3-Dimensional Sketch	Molecular Geometry & approx. bond angle (Actual angles differ due to lone pairs requiring more space.)	Example
5	0	5	sp^3d		trigonal bipyramidal bond angle = 90°, 120°	$SbCl_5$
4	1	5	sp^3d		See-saw shaped bond angle ≈ 90°, 120°, 180°	SF_4
3	2	5	sp^3d		T-shaped bond angle ≈ 90°, 180°	ClF_3
2	3	5	sp^3d		Linear bond angle = 180°	I_3^-
6	0	6	sp^3d^2		octahedral bond angle = 90°	SF_6
5	1	6	sp^3d^2		Square-pyramidal bond angle = 90°	$SbCl_5^{2-}$
4	2	6	sp^3d^2		Square planar bond angle = 90°	XeF_4

In the case of SF₄, the Lewis structure and geometry are shown below.

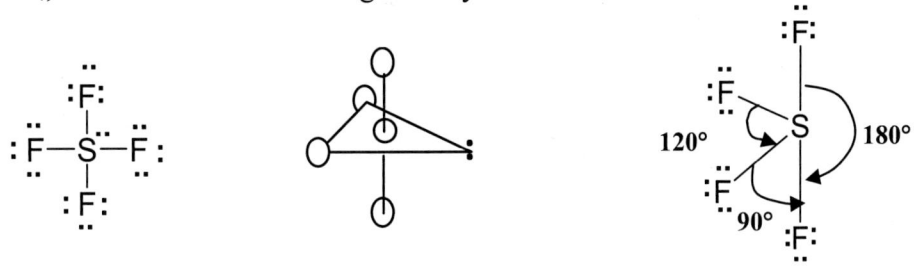

Fig. 11.3   Lewis Structure   3-D Arrangement   See-Saw
                                    of electron groups   Molecular Geometry

So far it is evident that the hybridization and shape and of a simple molecule with one central atom (as shown above for CO_2, BF_3, and CCl_4) can easily be determined. Larger molecules however, do not allow for such simple designations as linear, trigonal planar, tetrahedral, etc. because they do not have just one central atom. Any atom in a large or small molecule (or polyatomic ion) which has two or more atoms attached to it is considered a *"central"* atom. Hence larger molecules will have two or more central atoms. Take for example the hypothetical molecule shown below, it looks complicated, (it cannot be described using just one of the shapes in Table 11.1), but if you work your way through small portions at a time, you can identify the shapes about individual central atoms and figure out the corresponding hybridization of those central atoms.

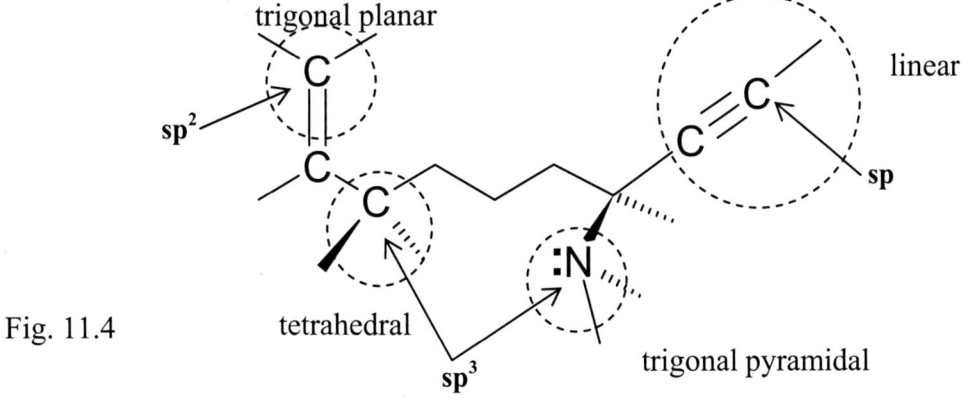

Fig. 11.4

It is useful to remember that the hybridization of a central atom is dependent on the total number of groups of electron on the central atom, regardless of how many bonds and how many of the electrons are bonding or nonbonding. The table below summarizes how one can predict the hybridization from just the total number of groups of electrons.

Table 11.2: Relationship between Groups of Electrons and Hybridization

Total # Groups of Electrons	Hybridization	Bond Angle
2	sp	180°
3	sp²	120°
4	sp³	109.5°
5	sp³d	90°, 120°, 180°
6	sp³d²	90°, 180°

For example, in the molecule shown on the right, the N atom has a total of three groups of electrons: one group bonding to H, one group bonding to C, and the lone pair. The N atom therefore has the sp^2 hybridization. The C atom has a total of 3 groups and it has the sp^2 hybridization. The O atom has 4 groups and it has the sp^3 hybridization. Note that the molecule is bent about the N, trigonal planar about the C, and bent about the O.

Molecular Polarity

The molecular geometry determines whether a molecule with polar bonds will be polar or nonpolar. Polarity of bonds results from the difference in electronegativity between atoms that share electrons. The greater the difference in electronegativity between bonded atoms, the greater the bond dipole. For molecules with two or more bond dipoles, the sum of all the dipoles or net dipole, called the dipole moment (μ), can be either zero or greater than zero. If a molecule is symmetrical in shape the bond dipoles will cancel and the dipole moment will be zero making the molecule nonpolar. Unsymmetrical molecules with polar bonds will always have a net dipole. In some cases the dipoles will re-enforce each other and in other cases there will be some cancellation but the molecule will have a net dipole moment (μ ≠ 0) and will be polar. Consider the following molecules:

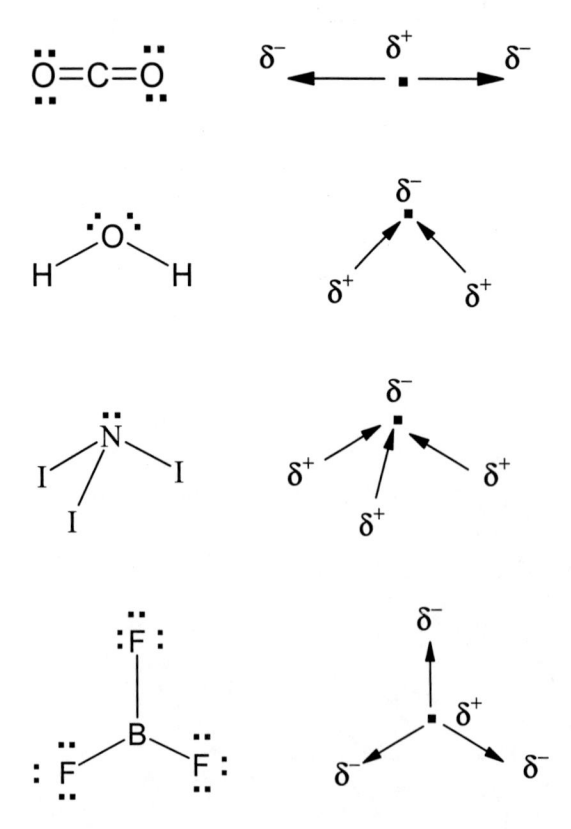

Although the C=O bonds are polar, because the bond dipoles are pulling in exactly opposite directions, they cancel out, and the CO_2 molecule is **nonpolar**. μ = 0.

In H_2O, the H–O bonds are polar, but this time the bond dipoles are at an angle and therefore they do not cancel out. The molecule is **polar**. μ > 0.

The three N-I bonds are polar but they are all "pushing" upwards and therefore do not cancel out. The molecule is **polar**. μ > 0.

The three B-F bonds are polar and the symmetry of the molecule allows the bond dipoles to cancel out. Thus the molecule is **nonpolar**. μ = 0.

Experimentally, polarity of a molecule is determined by placing a sample between charged plates and determining whether the molecules align themselves with the partial positive end pointing towards the negative plate (and the partial negative end pointing towards the positive plate). For example, a sample of water molecules would orient themselves in a charged field as shown in the diagram—that is, with the hydrogen atoms pointing towards the negatively charged plate and the oxygen atom pointing towards the positively charged plate. In the case of a polyatomic ion, the polarity is far less significant than the charge of the ion and therefore cannot be measured in this manner. Nonpolar molecules such as CO_2 and BF_3 will not sense the charged field and will be randomly distributed. In general, linear, trigonal planar, tetrahedral, trigonal bipyramidal, and octahedral molecules with identical atoms attached to the central atom are nonpolar and will be randomly distributed between charged plates in an electric field.

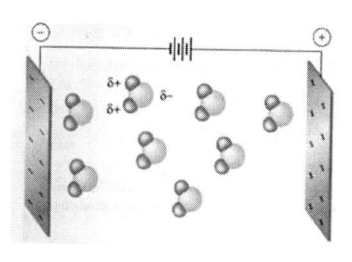

It is important to note that physical properties such as boiling point, solubility, and to some extent, the physical states are influenced by polarity. For example, H_2O is a liquid at room temperature and has a relatively high boiling point while nonpolar molecules of similar molar mass such as CH_4 and BH_3 are gases with much lower boiling points. The H_2O molecule exhibits these properties only because it is a polar molecule **and** is able to form relatively strong intermolecular bonds. If water was not such a highly polar molecule with extensive intermolecular attraction, it would be a gas at room temperature.

In this experiment you will see how we can predict the molecular polarity of a substance beginning with its molecular formula. Starting with a molecular formula you will determine the number of valence electrons available for bonding and work out a Lewis structure. From the number of groups of electrons in the Lewis structure you will predict the molecular geometry and hybridization about each central atom (as indicated in Table 11.1). You will then consider the difference in electronegativity of the atom at each end of a bond to establish the bond polarity. Examination of both the bond polarity and molecular geometry will then allow you to predict the molecular polarity.

Equipment/Materials:
Ball-and-Stick molecular model kit (wooden or plastic), Styrofoam balls, toothpicks

General Instructions: The balls are colored to allow you to distinguish amongst the different elements. Usually, the colors correspond in the following manner:
- black - any atom with three or four covalent bonds
- red - oxygen
- green - chlorine
- orange - bromine
- violet - iodine
- yellow - hydrogen (wooden)
- white - hydrogen (plastic) *(if available)*

138   EXPERIMENT 11: MOLECULAR GEOMETRY & POLARITY

Each rod represents a single covalent bond. A double bond is represented by two springs as shown, and a triple bond, by three springs (for the wooden model set); two longer flexible rods for double bonds, and a triple bond by three longer flexible rods (for the plastic model set). The Ball-and-Stick model does not work for cases where the Octet Rule has been violated. In such cases, you should use Styrofoam balls as atoms and toothpicks to show the spatial orientation of covalent bonds and lone pairs. A toothpick represents either one covalent bond or a lone pair (2 nonbonding electrons). You will only need to show the lone pairs on the <u>central</u> atom(s). For example, a model of BrO_2^- is shown below. The toothpicks connecting Br to O represent covalent bonds. (One of the oxygen atoms has two toothpicks to depict a double bond.) The other two toothpicks represent lone pairs. It is not necessary to show the lone pairs on the oxygen atoms.

Lewis structure   Model

:Ö=Br—Ö:⁻   (Molecular geometry is "bent," with a bond angle of around 109.5°)

For 5 groups of electrons, place three toothpicks on the same plane at 120° apart. These are said to be positioned at "equatorial" positions. The remaining two toothpicks are perpendicular to this plane, and are at "axial" positions. Note that lone pairs are always placed in the equatorial positions first.

For 6 groups of electrons, place four toothpicks on the same plane, at 90° to each other. The remaining two toothpicks are perpendicular to this plane.

Since all 6 positions are equivalent, there are no axial or equatorial positions.

Procedure:

1. Construct a data page with the following headings in your lab notebook. In order that you might have enough room to work, you should create the table sideways (i.e. landscape). You might need two pages. Record all information directly into your table.

Lewis Structure	Hybridization of Central Atom	Bond Angle	3-D Structure and Name of Shape (showing bond dipoles & net dipole)	Molecular Polarity

EXPERIMENT 11: MOLECULAR GEOMETRY & POLARITY

2. Draw a suitable Lewis structure for each of the following species:

   a) ClCN (Cl is chlorine, C is the central atom)
   b) SeCl$_2$
   c) NBr$_3$
   d) SCl$_6$
   e) AsF$_5$
   f) H$_2$CO
   g) ClO$_3^-$
   h) NO$_2^+$

   Have your instructor check your answers before you continue to the next steps.

3. Predict the hybridization of the central atom in each of the Lewis structures drawn in step 2 above.
4. Use the molecular models to construct a 3-D structure for each Lewis structure drawn in step 2 above.
5. Use the 3-D structures constructed in step 4 above along with information in the Table 11.1 to help you draw and name the shape of each species. You must use solid and broken wedges when showing atoms not in the plane of the paper.
6. Use the 3-D structures constructed in step 4 above to determine the polarity of each species. Draw each species and indicate the bond dipoles (using ⊢⟶, δ+ and δ-) and the direction of the net dipole, where applicable.
7. Make a model of the amino acid, glycine, whose Lewis structure is shown below.

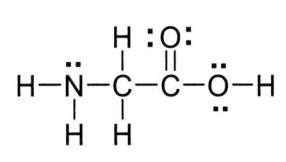

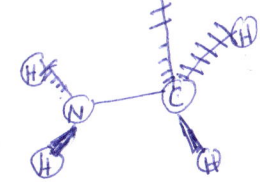

   ```
         H  :O:
         |  ||
    H-N-C-C-Ö-H
         |  
         H  H
   ```

   a. On a new page of your notebook, draw the geometric structure of glycine.
   b. Circle each of the four central atoms in glycine and indicate the hybridization of each central atom (as in Figure 11.4).
   c. Identify the spatial arrangement of atoms around each central atom (linear, trigonal planar, trigonal pyramidal, etc.) as in Figure 11.4.

Pre-Lab Exercise: (Answers are likely to be collected at the beginning of class.)

1. What is the total number of groups of valence electrons in each of the following?
   (a) ClO$_2^-$ ion (Cl is chlorine)
   (b) COS
   (c) H$_3$O$^+$

2. Identify each of the following as having ionic bonds, covalent bonds or both:
   (a) NaClO$_2$
   (b) COS
   (c) BeCl$_2$ (Cl is chlorine.)

3. Give a Lewis structure for each of the following. If resonance exists, give **all** the resonance structures as well:
   a) NI_3 (I is iodine)
   b) COS (C is the central atom)
   c) $ClBr_2^+$ (chlorine is the central atom)
   d) O_2
   e) N_2
   f) KBr

4. Determine which of the three resonance structures for ClF_2^+ is most stable and explain why. (Hint: Show the formal charges in these structures and think about where the charges are.)

:F̈—C̈l—F̈:   ⟷   :F̈=C̈l—F̈:   ⟷   :F̈—C̈l=F̈:

5. Name the molecular geometry and give the bond angles of the most stable structure in each of the following:
   a) NI_3 b) COS   c) $ClBr_2^+$

6. Give the hybridization of the central atom of each of the species in Question 5 above.

7. Using only a periodic table (without referring to the Electronegativity Table unless necessary) indicate the polarity of the bonds shown below. If the bond is nonpolar, say so:
   a) N–O
   b) C–O
   c) C–S
   d) Cl–I
   e) H–Cl
   f) H–K

8. Sketch each species to show correct molecular geometry, bond dipoles and then state whether each species is polar or nonpolar. If resonance structures exist, use the most stable one.
   a) NH_3 b) COS   c) $ClBr_2^+$

Post-Lab Questions:

1. If the lone pair were not present on the nitrogen in NBr_3 in (2c) of the procedure, what would be the shape of the species?

2. What would happen to the polarity if the lone pair were not present on the nitrogen in the NBr_3 molecule?

3. One of the Lewis structures of $C_4H_9NO_2$ is as shown:
   a) Identify the spatial arrangement of atoms about each central atom (linear, trigonal planar, tetrahedral etc.).
   b) Identify the hybridization of each central atom (sp, sp^2 etc.)
   c) Identify the likely bond angles at each central atom.
   d) Copy the Lewis structure onto your own paper and put in formal charges, if any.
   e) Give a resonance structure to this molecule.

Experiment 12: MOLAR VOLUME OF AN IDEAL GAS

Purpose: Determine the molar volume of a gas at standard temperature and pressure (STP, 0 °C and pressure of 1 atm)

Performance Goals:
- Collect and measure the volume of a gas using an eudiometer tube
- Make corrections to adjust for the difference in pressure inside and outside the eudiometer tube
- Make unit conversion from inHg to mmHg
- Make unit conversion from cmH_2O to mmHg
- Make corrections in the pressure of a gas collected over water
- Gain an appreciation for the practical use of gas laws and stoichiometry as applied to the experimental determination of the molar volume of an ideal gas

Introduction:

The term, *molar volume*, refers to the volume of one mole. Since volume varies with temperature and pressure, it would be meaningless to compare volumes measured under different conditions. It is therefore customary to make the measurements under convenient laboratory conditions and then convert the measurements to the volume at *standard temperature and pressure* (STP, 0°C and 1 atm).

For an ideal gas, the size of a gas particle (atom or molecule) is insignificant compared to the size of the container holding the gas, and there is no interparticle interaction (no attraction nor repulsion between particles). The molar volume of the gas at a particular temperature and pressure is independent of the <u>type</u> of gas.

In this experiment, hydrogen gas generated by the reaction of magnesium with hydrochloric acid is collected over water.

$$Mg(s) + 2HCl(aq) \longrightarrow MgCl_2(aq) + H_2(g)$$

The volume of the gas is measured and the number of moles of gas is calculated from the mass of the magnesium strip used. By dividing the volume by the number of moles we obtain the molar volume at the temperature and pressure at which the experiment is performed. In order to find the molar volume at STP, we apply the Ideal Gas Law:

$$PV = nRT$$
where P = the pressure of the gas
V = the volume of the gas
n = the number of moles of gas
R = gas constant
T = the temperature in K

142   EXPERIMENT 12: MOLAR VOLUME OF A GAS

From our experimental data we have V at the P and T under experimental conditions, and the question at hand is what V equals to when P is 1 atm and T is 0 °C. In this problem we have two sets of conditions: P_1, V_1, T_1 for the experimental conditions and P_2, V_2, T_2 for the STP conditions. R is a constant and in this case, so is n, which equals 1 mole. If we rearrange the Ideal Gas Law so that all the variables are on the left side of the equation, and the constants are on the right, we have

$$\frac{PV}{T} = nR = \text{constant}$$

This means $\frac{PV}{T}$ of 1 mole of gas at any condition must equal to the same constant, and therefore we arrive at the equation, $\frac{P_1 V_1}{T_1} = \frac{P_2 V_2}{T_2}$. We are looking for V_2, and so we rearrange the equation to solve for V_2:

$$V_2 = \frac{P_1 V_1 T_2}{T_1 P_2}$$

where V_2 = molar volume at STP
T_2 = 0 °C (converted to units of K)
P_2 = 1 atm
V_1 = the experimental molar volume
T_1 = the experimental temperature of the gas in K
P_1 = the experimental pressure of the gas in atm

Note: 0 °C and 1 atm for STP are exact numbers.

Equipment/Materials

Analytical balance (to 4 decimal places), eudiometer tube (50-mL), one or two-holed rubber stopper that fits on mouth of eudiometer tube, ruler, 800-mL or 1000-mL beaker, 50-mL beaker, piece of thread about 25 cm in length, magnesium strips of 4.5 cm long between 0.03–0.04 g each, 6 M HCl, deionized water

Procedure: *(Work with one partner.)*

CAUTION: WEAR SAFETY GOGGLES AT ALL TIMES. HANDLE THE 6 M HCl WITH CARE!

1. Obtain an eudiometer tube and an 800-mL beaker. Rinse the eudiometer tube three times with deionized water as you would with a buret.

2. For this experiment you must use the electronic balance that gives you 4 decimal places. If you don't know which one it is, ask your instructor. Record the mass of a strip of the metal approximately 4.5 cm long. Make sure it is between 0.0300 g to 0.0400 g and then record its mass to the nearest 0.0001 g. *__It is essential that the mass is within this range!__* Fold it into thirds and obtain a piece of string about 25 cm in length. Tie the end of the string to the metal strip.

3. Fill the 800-mL beaker about three-fourths full of tap water. Students with bigger hands may prefer to ask for an 1000-mL beaker.

4. With a 50-mL beaker, first pour about 15 mL of 6 M HCl into the eudiometer tube. Some of this acid will no doubt be adhering to the inside surface of the tube and introduce a source of error. With a wash bottle, wash down the sides of the tube with a couple of milliliters of deionized water. Next, carefully pour in about 15 mL of deionized water being careful not to disturb the acid on the bottom of the tube. Dangle the metal sample into the eudiometer tube, with part of the string hanging outside, then fill the eudiometer tube to the brim, with deionized water. The metal sample should be dangling at a depth of about 15 cm. You may use the temperature probe to help push the Mg strip down.

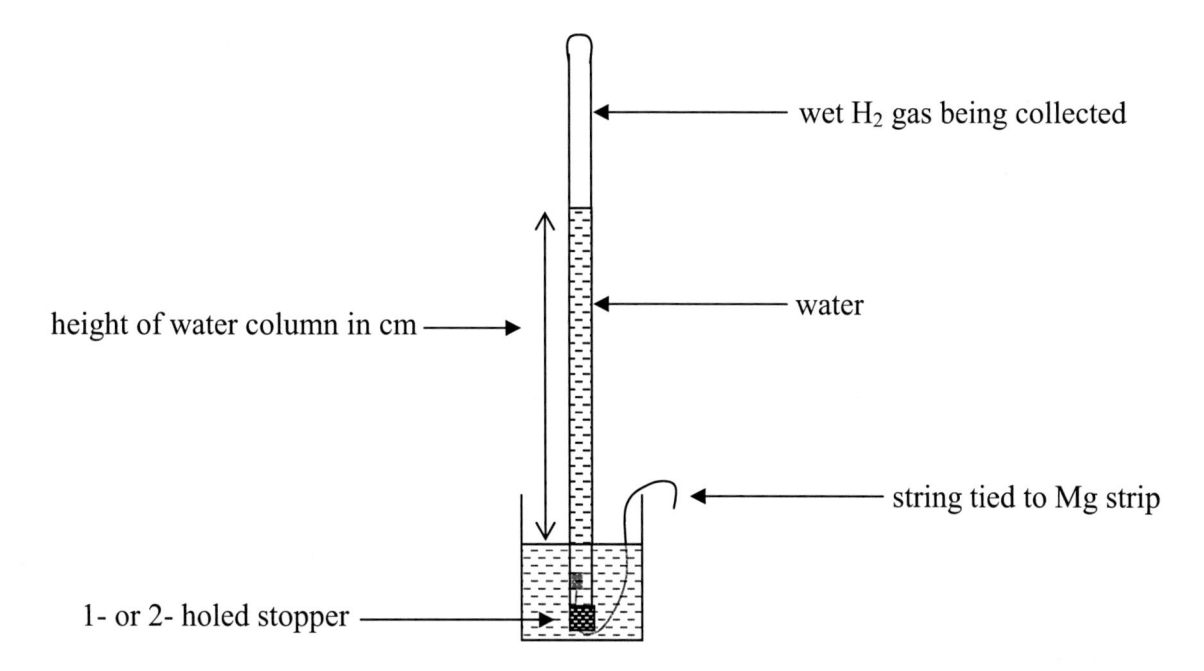

- wet H_2 gas being collected
- water
- height of water column in cm
- string tied to Mg strip
- 1- or 2- holed stopper

5. Hold the sample in place by inserting a one- or two-holed stopper into the opening of the tube. (*It is **not** necessary to thread the string through the hole.*) Hold your thumb tightly over the stopper, quickly invert and set it in the beaker of water (see Figure). Take care that you are not introducing any air bubbles into the tube.

6. Clamp the tube with its mouth about 1" below the surface of the water. The acid, being denser, will start to diffuse downwards, and you should be able to observe the evolution of gas when it comes in contact with the metal.

7. When the reaction is complete, examine the sides of the tube to make sure there is no unreacted metal clinging to the sides.

8. Measure the distance between the water levels inside and outside the eudiometer tube with a ruler and record in cm. Do not move the tube during this process.

EXPERIMENT 12: MOLAR VOLUME OF A GAS

9. <u>Without moving the tube</u>, record the volume of the gas by reading the graduations on the eudiometer tube, which is calibrated in units of mL. (If you moved the tube your volume reading would not correspond to the distance measurement in Step 8.)

10. Insert a temperature probe in the beaker so that its bulb is near the mouth of the tube. Allow it to equilibrate for a few minutes and record the temperature.

11. Record the barometric pressure. Your instructor will provide this in units of inHg. Convert this to mmHg *in your lab notebook* and check with your instructor to see that this is done correctly. You will waste a lot of time in your calculations if you have error at this early stage of your calculations.

12. Rinse out the beaker, eudiometer tube and string and repeat the experiment with another strip of magnesium. Use a new piece of string if you wish.

13. **CLEANUP:** Make sure no string or Mg metal ends up in the sink! Excess 6 M HCl must go into the special waste container in the hood. Other solutions can go down the drain.

Prepare your Data Table in your lab notebook:

	Trial #1	Trial #2
Mass of magnesium		
Volume of gas collected (in mL)		
Height of water column (in cm) (distance between water levels)		
Water temperature (in °C)		
Barometric pressure (in units of inHg)*		

*****Reference** *(State source of barometric pressure in the proper format)*
Show unit conversion calculation to convert the barometric pressure from inHg to mmHg here *(in lab notebook)*:

EXPERIMENT 12: MOLAR VOLUME OF A GAS

Sample Calculations:

When 0.0445 g of magnesium metal was allowed to react with excess hydrochloric acid, 46.50 mL of hydrogen gas was collected over water. The barometric pressure at the time was reported to be at 30.07 inHg. The temperature of the water near the mouth of the eudiometer tube was 26.4 °C. The water level inside the tube was measured to be 23.01 cm higher than the water level in the beaker. Calculate the molar volume of hydrogen at STP, and the percent error in this value.

$$Mg\ (s)\ +\ 2HCl\ (aq) \longrightarrow MgCl_2\ (aq)\ +\ H_2\ (g)$$

$$x\ mol\ Mg = 0.0445\ g\ Mg \left(\frac{1\ mol\ Mg}{24.31\ g\ Mg} \right) = 0.001830\ mol\ Mg$$

$$x\ mol\ H_2 = 0.001830\ mol\ Mg \left(\frac{1\ mol\ H_2}{1\ mol\ Mg} \right) = 0.001830\ mol\ H_2 \quad \text{(Based on balanced equation above)}$$

$$\text{Molar volume of } H_2 \text{ at experimental conditions} = \frac{0.04650\ L}{0.001830\ mol} = 25.40\ L\ mol^{-1}$$

$$\text{Barometric pressure in units of mmHg} = 30.07\ inHg \left(\frac{2.54\ cmHg}{1\ inHg} \right) \left(\frac{10\ mmHg}{1\ cmHg} \right) = 763.77\ mmHg$$

The water level inside the tube is higher than the water level outside the tube (water level in the beaker) because the pressure inside the tube is less than that pressure outside (the barometric pressure).

To adjust for the difference in height of water level,

$$\text{Pressure inside the tube} = (\text{Pressure outside the tube} - \text{Height of water column})$$

Conversion of height of water column from cmH$_2$O to mmHg

$$= 23.01\ cmH_2O \left(\frac{10\ mmH_2O}{1\ cmH_2O} \right) \left(\frac{1\ mmHg}{13.5\ mmH_2O} \right) = 17.04\ mmHg$$

Pressure inside the tube = Pressure of wet gas
= Barometric Pressure − Height of water column
= 763.77 mmHg − 17.04 mmHg = 746.73 mmHg

Pressure of dry gas = Pressure of wet gas − Vapor Pressure of Water at 26.4°C*
(*from Table 12.1)

$$= 746.73\ mmHg - 25.81\ mmHg = 720.92\ mmHg$$

Experimental conditions for 1 mol of gas | STP conditions for 1 mol of gas
$P_1 = 720.92$ mmHg | $P_2 = 760$ mmHg (exact number)
$T_1 = (26.4 + 273.15)\ K = 299.55\ K$ | $T_2 = (0°C + 273.15)\ K = 273.15\ K$ (2 decimal places)
$V_1 = 25.40$ L | $V_2 = ?$ L

$$\text{Molar Volume at STP} = V_2 = \frac{P_1 V_1 T_2}{T_1 P_2} = \frac{(720.92\ mmHg)(25.40\ L)(273.15\ K)}{(299.55\ K)(760\ mmHg)}$$

$$= 21.97\ L = 22.0\ L$$

Commonly Accepted Molar Volume of an Ideal Gas = 22.4 L/mol

Error = 22.0 − 22.4 L/mol = −0.4 L/mol or Error = | −0.4 L/mol | = 0.4 L/mol

$$\text{Percent Error} = \frac{-0.4\ L/mol}{22.4\ L/mol} \times 100 = -2\ \%\ \text{Error} \qquad \%\ \text{Error} = \frac{0.4\ L/mol}{22.4\ L/mol} \times 100 = 2\ \%\ \text{Error}$$

146   EXPERIMENT 12: MOLAR VOLUME OF A GAS

Prelab Exercise:
Your instructor will specify whether you are to submit the answers to these questions or take pre-lab quiz with similar questions.

1. How many moles of gas are produced if 0.037 g of magnesium reacted? (Watch your significant figures! Hint: How many sig. fig. are in 0.037 g?)

    $$Mg\ (s)\ +\ 2HCl\ (aq)\ \longrightarrow\ MgCl_2\ (aq)\ +\ H_2\ (g)$$

2. Convert 26.8°C into units of Kelvin. Show your setup and watch your sig. fig. Consult your textbook for this conversion.

3. Convert 689 mmHg into units of atm. Show your setup and watch your sig. fig. Consult your textbook for this conversion. Note: 1 atm = 760 mm Hg where 760 is an exact number (not 2 sig. fig.)

4. If the volume of a gas is 25.8 mL at 28.7°C and 728 mmHg, what is its volume at 15.2°C and 757 mmHg? You should be able to do this calculation by reading the experimental write-up. (Hint: Temperature must be converted to degrees kelvin.)

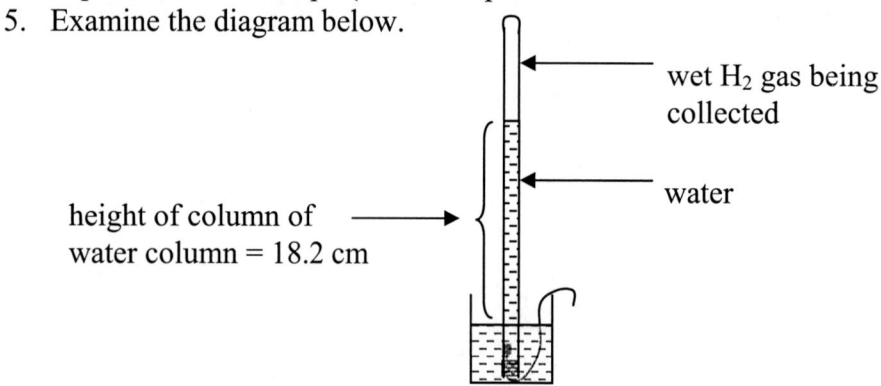

5. Examine the diagram below.

    → wet H_2 gas being collected

    → water

    height of column of water column = 18.2 cm

    a) The height of the water column is 18.2 cmH₂O. Convert this to units of mmHg using dimensional analysis.
    (1 mmHg is equivalent to 13.5 mmH₂O. Note: 13.5 has 3 sig. fig.)

    b) If the barometric pressure is 762.3 mmHg, what is the pressure of the wet H_2 gas in mmHg?

6. If the pressure of the wet H_2 gas is 580.4 mmHg and the temperature of the water is 21.2°C, what is the pressure of the dry gas in mmHg, and in atm? Show your work.

Post-Lab Questions: (Please type your answers.)

1. The reaction in this experiment is an oxidation-reduction reaction. What is being oxidized and what is being reduced? Explain your answer.

2. In order for this experiment to work properly to give the correct molar volume of an ideal gas, which must be the limiting reactant in the reaction? Explain your answer.

3. If there is unreacted metal remaining at the end of the experiment, what would that do to your calculated molar volume? Be specific. Do not merely say that "it would be

incorrect." Would the calculated molar volume be too small or too large? Explain your answer.

4. If you had used 0.030 g of magnesium in the first trial and 0.040 g in the second, would you expect the molar volume at STP to be larger or smaller in the second trial? Explain your answer. Do not merely explain by saying molar volume is an intensive property. Explain WHY its molar volume is not dependent on the sample size.

5. Hydrogen is a real gas.
    a) Under what conditions do you expect it to deviate from ideal gas behavior? Explain.
    b) Is it reasonable to consider hydrogen as an ideal gas under the conditions of this experiment? Explain.

148 EXPERIMENT 12: MOLAR VOLUME OF A GAS

HOW TO READ A MERCURY MANOMETER:

In this experiment you have to deal with the height of a column of water inside the eudiometer and relate it to the pressure of the gas trapped inside. You will understand how that works by studying the discussion below as to how a mercury manometer works.

A *manometer* is an instrument that gives us the pressure of a gas (not to be confused with a *barometer* which gives us the pressure of the atmosphere around us).

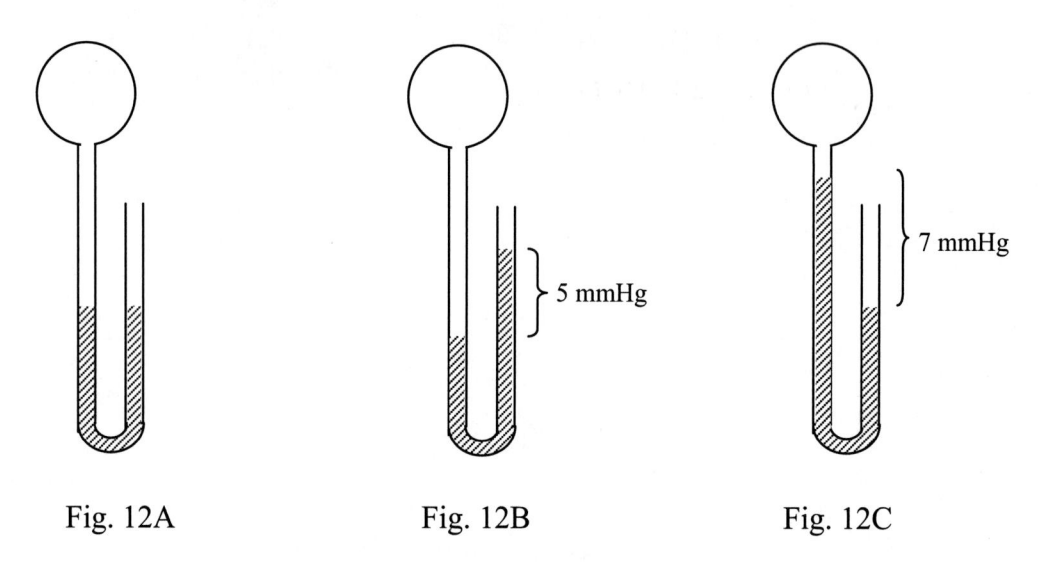

Fig. 12A   Fig. 12B   Fig. 12C

In the figures above we see a gas sample inside the bulb connected to a U-tube filled with mercury. One end of the U-tube is opened to the atmosphere.

In Fig. 12A, the levels of the mercury are even. This indicates that the pressure of the gas P_{gas} is equal to the pressure outside which can be obtained by reading a barometer which gives us the atmosphere pressure (P_{atm}).
$$P_{gas} = P_{atm}$$

In Fig. 12B, the level of the mercury column on the left is lower than that on the right. This indicates that the pressure of the gas is <u>higher</u> than the atmospheric pressure.
$$P_{gas} > P_{atm}$$
To be specific, it is higher by the difference in height of the two columns measured in mm. For example, if the distance between the two levels were 5 mm, we can say
$$P_{gas} = P_{atm} + 5 \text{ mmHg}$$

In Fig. 12C, the level of the mercury column on the left is higher than that on the right. This indicates that the pressure of the gas is <u>lower</u> than the atmospheric pressure.
$$P_{gas} < P_{atm}$$
If the distance between the two levels were 5 mm, we can say
$$P_{gas} = P_{atm} - 7 \text{ mmHg}$$

Since these measurements were done on mercury rather than water, the unit is mmHg.

Review Information for CHEM 122 Lab Practical Final Exam

Check your lab schedule for the exact date of your final examination for CHEM 122. It is not during "Final Exam Week" that the college reserves for the lecture courses, but the week before. The exam will count about of the student's grade in CHEM 122. Students will be expected to make measurements as well as perform calculations similar to those made in the experiments during the semester.

During the week of the laboratory final, only the exam will be given. Students will not be able to make up past lab experiments when the practical is set up in the lab. Only students taking the exam will be allowed in the lab.

Students should bring a **non-programmable** scientific calculator, pen and pencil to the exam and nothing else. **This is a closed-book exam and you will be working individually**. Cells phones must be turned off and cannot be used as a calculator.

HOW TO STUDY: Go through each experiment and think about how each measurement was made, then take the data from your Data Table and review all the calculations that were done without the aid of the lab manual. Also go through all the pre-lab quizzes, questions provided to help you prepare for these quizzes, sample calculations and post-lab questions.

The exam will test the student's ability to do the following:

1. Perform measurements to the proper number of significant figures on the following apparatus: balances, graduated cylinders, rulers, thermometers, and burets.

2. Be able to distinguish amongst the three types of error: operator, instrumental and method error. Given the description of an experiment, be able to give examples of each of the three types of errors for the experiment.

3. Be able to plot a linear graph by hand, from a given set of data, give the acceptable scales and axis titles, and obtain information such as the slope, y-intercept, equation for the line and predicted data from extrapolation and interpolation.

4. Identify, by correct name and spelling, the apparatus used in the experiments throughout the semester. (Some, but not all, are shown on p. 5.)

5. Determine the density of a solid or liquid by making measurements with the use of the appropriate apparatus and performing the necessary calculations.

6. Recognize and be able to explain the difference between absolute error and relative error.

7. Write molecular, total ionic and net ionic equations for chemical reactions of the type performed in the "Chemical Reactions" experiment.

8. Make measurements and calculations of the type used in the "Chromatography", "Composition of a Hydrate", "Acid-Base Titrations", and the "Molar Volume of an Ideal Gas" experiments.

> - You will <u>not</u> be prompted as to which measurements must be made to arrive at an answer.
> - You will <u>not</u> be guided through the calculations step by step.

9. Determine the limiting reactant, theoretical yield and percentage yield for a reaction similar to that performed in the "Synthesis of Tris(ethylenediamine)nickel(II) Chloride" experiment.

10. Calculate the heat transfer of the type found in the "Calorimetry" experiment from data provided.

11. Collect data to prepare a calibration graph and then make use of the graph.

12. Be able to calculate the wavelength of a transition of an electron in a hydrogen atom such as in the "Atomic Spectroscopy" experiment, and given a trendline equation of a calibration curve be able to make use of it to correct for an inaccurate instrument.

13. Given a molecular formula be able to draw its Lewis structure, make a 3-dimensional sketch of its molecular geometry, name the geometry, identify its state of hybridization, and determine whether it is polar or nonpolar.

APPENDICES

Appendix 1: REVIEW ON SIGNIFICANT FIGURES

Significant figures are critical in the lab because they indicate how precise the measurements are. For example, if you were conducting a clinical trial for a new drug, you would need to calculate how different the results were in patients taking the drug. If you do not know how precise the data is (i.e. how many significant figures), then you will not know how reliable the results are. Below is a summary of the rules that are to be followed not only when recording the values but also when performing mathematical manipulations.

Part A. Recording Data to the Appropriate Number of Significant Figures: The best rule of thumb is that you record to one-tenth of the smallest increment shown on the apparatus (one digit beyond what you can easily read). For example the graduated cylinder shown in Figure 1A has nine lines between 30 mL and 40 mL, (divided into 10 divisions) so each division represents 1 mL. One-tenth of 1 mL is 0.1 mL. This means all readings on this graduated cylinder should be recorded to one decimal place. In the figure below, the correct reading would be 32.5 or 32.6 mL. There is always going to be uncertainty in the last digit.

In Figure 1B, the smallest division is 0.1 mL. One-tenth of 0.1 mL is 0.01 mL. All readings with this graduated cylinder should be recorded to two decimal places. The correct reading would be 8.47 mL or 8.48 mL. It is understood that there is always uncertainty in the last digit of any measurement.

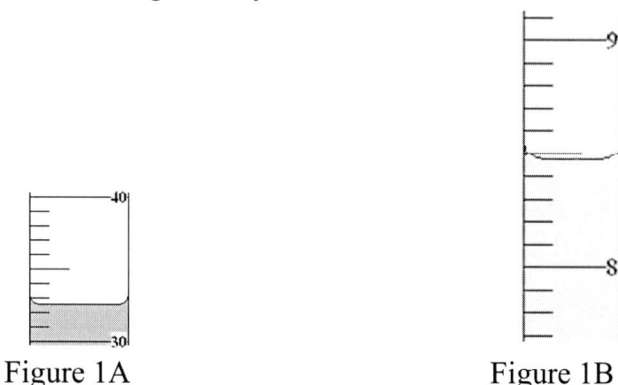

Figure 1A   Figure 1B

Once you have recorded the data, you then follow a set of rules to ensure that you keep the same level of precision in your final answer as you had in your recorded data. The rules are there so that you do not end up with more precision than is allowed by the apparatus just by doing calculations.

<u>All digits are significant except for two cases:</u>
   1) Leading zeroes (zeroes to the left of the first non-zero digit) are NOT significant because they merely hold the decimal place.
   0.00<u>7</u> has one significant figure
   0.000<u>620014</u> has six significant figures
   <u>12.1231</u> has six significant figures.
   <u>1203.03</u> has six significant figures.

2) **Trailing zeroes (zeroes on the end of a number) in a number WITHOUT a decimal point <u>are ambiguous</u>.** They are <u>*assumed*</u> to be NOT significant. The ambiguity is removed by using scientific notation.

(Trailing zeroes in a number WITH a decimal point and zeroes between non-zero digits <u>are significant</u>.)

0.00<u>73500</u> has five significant figures.
<u>73500.00</u> has seven significant figures.
0.0<u>7305</u> has four significant figures.
<u>7350.</u> has four significant figures. The trailing zero is significant because of the decimal point.
73500 has ambiguity because it could have three, four or five significant figures. *It is <u>assumed</u>* to have only 3 significant figures. The trailing zeroes are <u>*assumed*</u> to be not significant.
   If it were to have 5 sig. fig., it should be written as <u>7.3500</u> x 10^4.
   If it were to have 4 sig. fig., it should be written as <u>7.350</u> x 10^4.
   If it were to have 3 sig. fig., it should be written as <u>7.35</u> x 10^4.

Part B. Treatment of Significant Figures During Calculations: What happens to these significant figures during calculations? The reported answer should have the same precision as that of the least precise number. To do so, you must follow these rules:

ADDITION & SUBTRACTION: When adding and subtracting numbers, line up the decimal places and report the number with the same number of **decimal places** as that with the least decimal places.

```
    1123.123      (3 decimal places)          1123.123      (3 decimal places)
 +     0.002123   (6 decimal places)        − 1120.1        (1 decimal places)
    1123.125123   (3 decimal places)             3.023      (1 decimal place)

Ans. 1123.125                               Ans. 3.0
```

If the numbers are in exponential form, they must first be adjusted to the same power before lining up the decimal place for addition or subtraction.

2.431 x 10^{12} + 0.001 x 10^{13} = ?

Incorrect to answer in 3 decimal places
  2.431   x 10^{12} ← *Exponents cannot*
+ 0.001   x 10^{13} *be different when*
= 2.441   x 10^{12} *you add.*

Correct to answer in 2 decimal places
  2.431 x 10^{12} ← *Convert to*
+ 0.01   x 10^{12} *same exponent*
  2.44   x 10^{12} *before adding.*

MULTIPLICATION & DIVISION: When multiplying or dividing, report your answer with the same number of **significant figures** as that with the smallest number of significant figures.

1123.123 x 0.0000123 x 1.1   = 0.01519585419
(7 sig.fig.) (3 sig.fig.) (2 sig.fig.)

Answer should have 2 sig. fig. = 0.015

Part C. Rounding Off: When an answer needs to be expressed with fewer significant figures, if the first digit to be dropped is >5, round up. If it is < 5, merely drop the remaining digits. If it is exactly 5 then it depends on the number immediately to its left. Round up if the digit to its left is odd, and truncate if the digit to its left is even.

The following numbers are each rounded to three significant figures.

Original Number	1.23124	0.013968	1.6$\underline{7}$5	0.00032$\underline{4}$5
Rounded Number	1.23	0.0140	1.68	0.000324

(odd → 1.6$\underline{7}$5; even → 0.00032$\underline{4}$5)

Part D. Exact Numbers: Counting and Definitions – When counting (not measuring) a number, the value that is obtained is considered to be *exact* and therefore has an infinite number of significant figures. For example, if there were 3 people in a room there is no uncertainty, thus there would be exactly 3.00000000… people in the room. Definitions are also an exact number. It is defined that there are 1000 milliliters in 1 liter, thus one can say there are 1000.00000000… mL in 1 liter. Constants like Avogadro's number are not defined, but instead have been calculated and therefore they do have a correct level of significance that you should be aware of. Exact numbers do not affect the number of significant figures during calculations.

Part E. Scientific Notation: Often numbers are so large, or so small, that it becomes quite cumbersome to express the numbers without the use of scientific notation. For example the speed of light in a vacuum is nearly thirty million meters per second. This value expressed in conventional notation would be 300,000,000 m/s where one needs to write 8 zeroes and one wonders how many of those are significant. Better stated using scientific notation the number becomes 3.00×10^8 m/s to the precision of 3 significant figures or 2.9979×10^8 m/s to the precision of 5 significant figures.

Scientific notation is a method of expressing a value, such that the number has only one non-zero digit to the left of the decimal place, followed by the appropriate number of significant figures to the right of the decimal place and then multiplied by 10 raised to an exponent expressing the order of magnitude of the number. The number 3021.1 has too many digits to the left of the decimal point. In scientific notation it should be written as 3.0211×10^3.

The number 0.025×10^7 is not in scientific notation because the digit to the left of the decimal point is zero. It should be expressed as 2.5×10^5 instead.

A-4   APPENDIX 1: REVIEW ON SIGNIFICANT FIGURES

Part F. Calculation of Average: When determining the average of several values, the average cannot end up with more precision than the values themselves. Thus, the average of 24.6 g and 24.7 g should not be recorded as 24.65 g but should be 24.7 g (rounded to one decimal place).

Part G. Significant Figures Obtained from Graphs: Values read off a graph or calculated from the trendline of the graph should *not be more precise* than the data used to create the graph. In addition, scales chosen for the graph should be such that it does *not yield answers that are less precise* than the data.

Part H. Keeping Track of Units: When recording measurements, **ALWAYS** include the units. For example if you are recording the reaction time as being 8, we would be wondering whether you mean 8 seconds, minutes or hours.

In showing your calculation setups, it is essential that you keep track of your units from beginning to end. One reason is it helps you catch careless algebraic mistakes. For example, if you were calculating the volume of a sample from its density and mass and set it up as shown below, you would see that the units do not work out properly: Volume cannot have a unit of 1/mL. The unit tells you that your setup is upside down.

Incorrect: $$\text{Volume} = \frac{1.09 \text{ g/mL}}{2.58 \text{ g}} = 4.22 \frac{1}{\text{mL}}$$

Correct: $$\text{Volume} = \frac{2.58 \text{ g}}{1.09 \text{ g/mL}} = 2.37 \text{ g} \times \frac{\text{mL}}{\text{g}} = 2.37 \text{ mL}$$

In recent publications, units in the denominator are expressed with exponents of -1, rather than with the use of a slash to avoid confusion. For example g/mol is written as g mol^{-1}. The exponents of the units should be treated as you would with the exponents of numbers. For example, (g/mL•mL) is confusing as written. Does it mean [(g/mL)•mL] or [g/(mL•ml)]? To avoid such confusion, [(g/mL) •mL], such as dividing density by the volume, is best written as (g mL^{-1} mL), which simplifies to (g):

$$(\text{g mL}^{-1} \text{ mL}) = \text{g mL}^{-1} \text{ mL}^1 = \text{g mL}^{-1+1} = \text{g mL}^0 = \text{g }(1) = \text{g}$$
Reminder: Anything to the power of zero equals one. (e.g. $x^0 = 1$)

There is another important reason why keeping track of units at every step is important. If you were to calculate the pressure of a gas using the ideal gas law, PV = nRT and you neglected to pay attention to units, you are likely to end up with the wrong answer: What is the pressure in torrs of 0.200 mole of gas occupying 962 mL at 25.0°C?

Incorrect: $$P = \frac{nRT}{V} = \frac{(0.200)(0.08206)(25.0)}{(962)} = 4.27 \times 10^{-4} \text{ torr}$$

Why is it incorrect? By including units you would have seen that the units do not cancel properly. T must be in units of K and V must be in L.

Correct: $$P = \frac{nRT}{V} = \frac{(0.200 \text{ mol})(0.08206 \text{ atm.L.mol}^{-1}\text{K}^{-1})(298 \text{ K})}{(0.962 \text{ L})} = 5.08 \text{ atm}$$

$$P = 5.08 \text{ atm}\left(\frac{760 \text{ torr}}{1 \text{ atm}}\right) = 3.86 \times 10^3 \text{ torr}$$

Appendix 2: PREPARATION & INTERPRETATION OF GRAPHS

All of you should have had some experience in plotting graphs. Some of you may have done this in the distant past. Some may have done it only in math courses where numbers are exact and significant figures were not considered. This appendix will serve as a review, as well as point out features of graphs used in science of which you may not be aware.

A graph is most commonly drawn to show a relationship between two quantities. The most common way in which data are plotted on a graph is in Cartesian coordinates with two axes drawn at right angles, and one quantity is plotted on the horizontal axis (the abscissa or the *x*-axis) and the other on the vertical axis (the ordinate or the *y*-axis).

A set of points on a graph is not itself especially useful. On the other hand, a straight line or smooth curve connecting the points can be very useful in showing a trend and whether there are significant outliers in the data. Such a straight line or curve allows one to *interpolate*, which is to find values between the measured quantities. When the curve is a straight line it is often possible and very useful to *extrapolate*, which is to predict values beyond the ones measured. Keep in mind, though, that the further one extrapolates from the measured values, the less reliable it becomes.

SELECTION OF THE SCALES:

1. SELECTION OF X-AXIS AND Y-AXIS: When we specify "A *versus* B," by convention, "A" goes on the *y*-axis and "B" goes on the *x*-axis. (The *y*-axis is the vertical axis and the *x*-axis is the horizontal axis.)

   A *versus* B

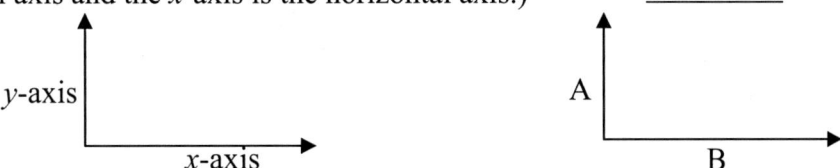

   When you are asked to plot a graph, read the assignment carefully to determine which data go on the *x*-axis and which go on the *y*-axis.

2. SELECTION OF A RANGE FOR THE SCALE: First of all, not all scales start with zero. If all of your data are in the range of 300 to 500, there is no reason to start with zero. The exception is if one needs to find the y-intercept (the *y*-value when *x* equals zero) on the graph, then obviously one will need to have *x* equals zero showing on the graph. Choose a scale with a range that goes from a number smaller than any of the numbers in the set of data, up to a number larger than any of the numbers in the set of data. In other words, the scale should not leave out any of points in the data.

   Second of all, keep in mind that the scale one selects should be **easily read**. In other words, it should be labeled at regular, convenient intervals. Figure A is an example of a poorly selected scale. Figure B is an example of a better scale.

A-5

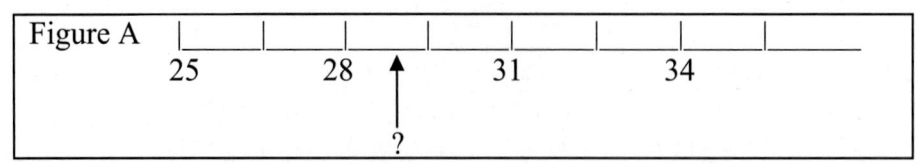

A-6   APPENDIX 2: PREPARATION & INTERPRETATION OF GRAPHS

```
Figure A  |____|____|____|____|____|____|____|____
            25        28  ↑     31        34
                          |
                          ?
```

In Figure A, although the numbers are evenly spaced, it would be difficult to read where the arrow is located. One does not want 2 squares to equal to 3 units apart (25 to 28).

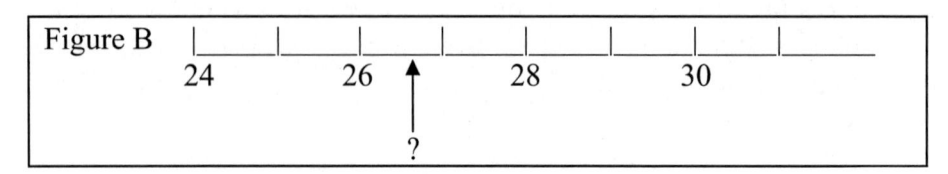

```
Figure B  |____|____|____|____|____|____|____|____
            24        26  ↑     28        30
                          |
                          ?
```

In Figure B, we can easily read the location of the arrow as being at 26.6 or 26.7.
Common increments to use on a scale are as follows:
0, 1, 2, 3, 4… or 0, 2, 4, 6, 8… or 0, 5, 10, 15, 20… or 0, 10, 20, 30, 40…
Avoid using increments such as 3, 6, 9, 12…or 4, 8, 12, 16… because we do not usually count in three's and four's.

3. SELECTION OF A SCALE SUCH THAT POINTS ARE NOT BUNCHED UP: Select a scale for each axis such that the points are spread out over the whole graph (not clumped up together leaving large blank spaces in the graph). A larger graph is preferred because they permit you to locate points with greater accuracy and precision.

Figure C shows an undesirable graph where points are clumped together. Figure D shows a graph with a more appropriate scale for the *x*-axis. **However, do not do this at the expense of ending with a scale that is too difficult to read as discussed above in Figure A.**

Figure C showing unacceptable x-scale   Figure D showing an acceptable x-scale

EXERCISE #1 ON SELECTING APPROPRIATE SCALES:
If your x-data are as follows, make a rough sketch of the *x*-axis, showing what you should have appear on the axis as minimum number, maximum number, other scale numbers and the minor tick marks between the numbers (as in Fig. B): 45.3 g, 58.4 g, 71.0 g, 82.6 g
Hint: You would not start with 0 g unless the y-intercept is needed.

(Answer to Exercise #1 is shown on page A-4)

APPENDIX 2: PREPARATION & INTERPRETATION OF GRAPHS   A-7

EXERCISE #2 ON SELECTING APPROPRIATE SCALES:
If your *x*-data are as follows, make a rough sketch of the *x*-axis, showing what you should have appear on the axis as minimum number, maximum number, other scale numbers and the minor tick marks between the numbers (as in Fig. B):
0.35 g, 0.58 g, 0.71 g, 0.82 g

EXERCISE #3 ON SELECTING APPROPRIATE SCALES:
If your *x*-data are as follows, make a rough sketch of the *x*-axis, showing what you should have appear on the axis as minimum number, maximum number, other scale numbers and the minor tick marks between the numbers (as in Fig. B):
1245 g, 1357 g, 1583 g, 1782 g

4. Always plot the points with a *__very sharp pencil__*. Mark each point with an X with the intersection of the X at precisely where your point should be. Do *__not__* mark the point with a big fat dot. One loses a lot of precision and accuracy as one cannot tell which part of the dot is precisely the point. If you wish you may circle the X to make it more visible.

5. In this class we will be dealing with only *linear graphs*. After you have plotted all the points, draw the best straight line (*__one__* line) through the points *__using a ruler and a very sharp pencil__*. If not all the points lie on the line, select a line such that it goes through most of the points, and with as many points above as there are points below the line. Extend this line all the way to the *y*-axis. **Note that not all lines go through the origin.** If one point seems way off, it is most likely that you misread the scale. It may also be due to an experimental error in your measurement. Consult with your instructor.

6. **Label each axis** with a title, including the units used. **Label the graph** with a title at the top of the page. An example is shown below:

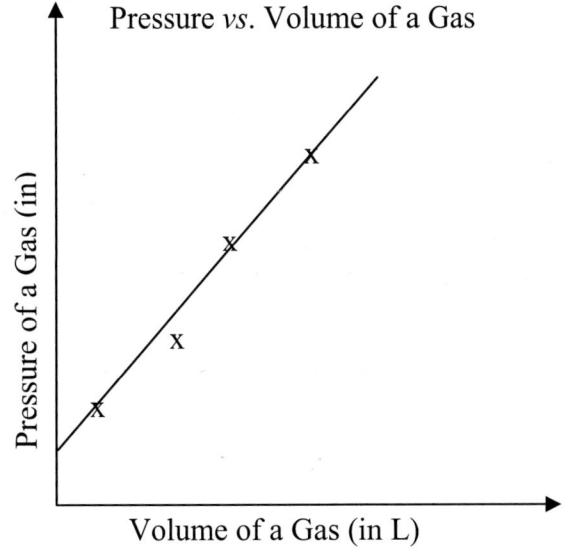

Pressure *vs.* Volume of a Gas

Pressure of a Gas (in)

Fig. E   Volume of a Gas (in L)

APPENDIX 2: PREPARATION & INTERPRETATION OF GRAPHS

7. DETERMINATION OF THE SLOPE:

The slope of the line is calculated from the coordinates of two points taken from the line. It is commonly referred to as "rise over run."

If we use the two points (x_1, y_1) and (x_2, y_2), the slope would be calculated as follows: *(By convention, we list the coordinates of a point by the x-value first, and then the y-value.)*

$$\text{slope} = \frac{\Delta y}{\Delta x} = \frac{y_1 - y_2}{x_1 - x_2}$$

In science, the *x*- and *y*-values have units. For the graph shown in Fig. E, the *x*-values have units of L and *y*-values have units of atm. Let's take the example of finding the slope from these two points: (3.15 L, 1.04 atm) and (6.82 L, 1.37 atm)
The slope would be calculated as follows:

$$\text{slope} = \frac{1.04 \text{ atm} - 1.37 \text{ atm}}{3.15 \text{ L} - 6.82 \text{ L}} = \frac{-0.33 \text{ atm}}{-3.67 \text{ L}} = 0.090 \text{ atm/L}$$

Note that units are included in the setup and in the final answer. Note also how significant figures are treated in the calculations.

WHICH POINTS SHOULD WE PICK FOR THE CALCULATION OF THE SLOPE?

a) Select two points that lie ***exactly*** on the line ***and*** are easy to read. The common mistake is to just pick two data points for the calculation. This is all right if the data points happen to lie ***exactly*** on the line. If they do not then all you would be doing is finding the slope of the line that joins only those two points. It would be much easier to find two points that lie at the cross hairs of a vertical and a horizontal line of the graph paper (so that you don't have to estimate between the lines).

b) You should also pick two points as far apart on the line as possible. If you pick two points right next to each other, you would not have enough significant figures when you calculate Δx or Δy.

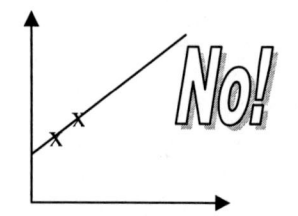

Points selected are too close to each other.
Fig. F

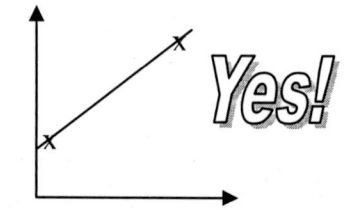

Points selected are appropriately far from each other, giving more sig. fig. in the slope.
Fig. G

8. DETERMINATION OF THE Y-INTERCEPT:

Conceptually, the term, *y-intercept*, refers to where the line intersects with the *y*-axis. Technically, it is the *y*-value when *x* equals zero. Since it is a **y-value**, the *y*-intercept should have the same units and decimal places as the *y*-data. In the graph shown in Fig. E, the *y*-intercept would have the units of atm.

ANSWER TO EXERCISE #1 ON SELECTING APPROPRIATE SCALES:
If your *x*-data are as follows, make a rough sketch of the *x*-axis, showing what you should have appear on the axis as minimum number, maximum number, other scale numbers and the minor tick marks between the numbers (as in Fig. B): *Hint: You would not start with 0 g.*
45.3 g, 58.4 g, 71.0 g, 82.6 g

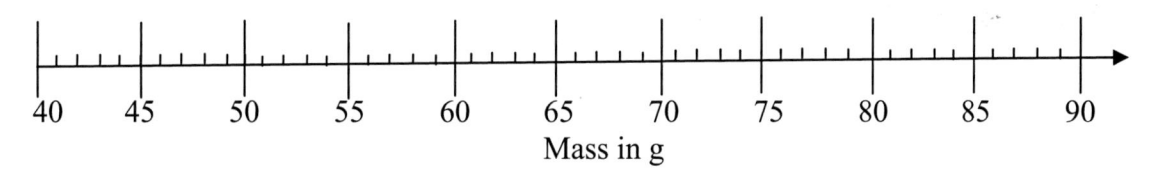

40 45 50 55 60 65 70 75 80 85 90
Mass in g

ANSWER TO EXERCISE #2 ON SELECTING APPROPRIATE SCALES:
If your *x*-data are as follows, make a rough sketch of the *x*-axis, showing what you should have appear on the axis as minimum number, maximum number, other scale numbers and the minor tick marks between the numbers (as in Fig. B):
0.35 g, 0.58 g, 0.71 g, 0.82 g

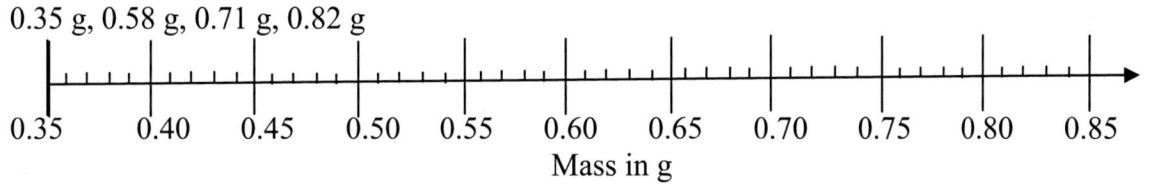

0.35 0.40 0.45 0.50 0.55 0.60 0.65 0.70 0.75 0.80 0.85
Mass in g

ANSWER to EXERCISE #3 ON SELECTING APPROPRIATE SCALES:
If your *x*-data are as follows, make a rough sketch of the *x*-axis, showing what you should have appear on the axis as minimum number, maximum number, other scale numbers and the minor tick marks between the numbers (as in Fig. B):
1245 g, 1357 g, 1583 g, 1782 g

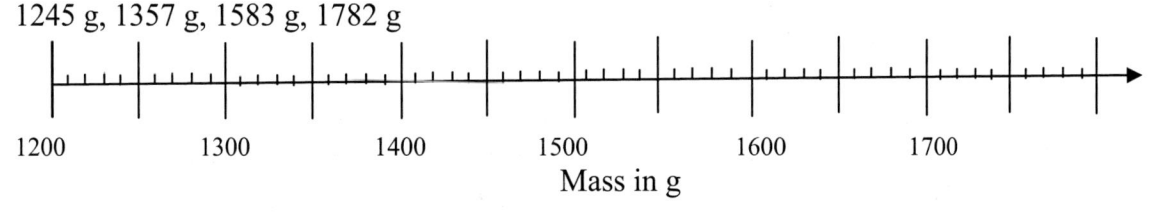

1200 1300 1400 1500 1600 1700
Mass in g

A-10   APPENDIX 2: PREPARATION & INTERPRETATION OF GRAPHS

EXERCISE #4: Answer the following questions based on the graph shown below. Watch your sig. fig. and units. *Write down your answers before you check the correct answer shown at the bottom of the page.*

a) What is the slope? First write down the coordinates of two convenient points you are going to use to calculate this slope.
b) What is the y-intercept?
c) What is the equation for the line?

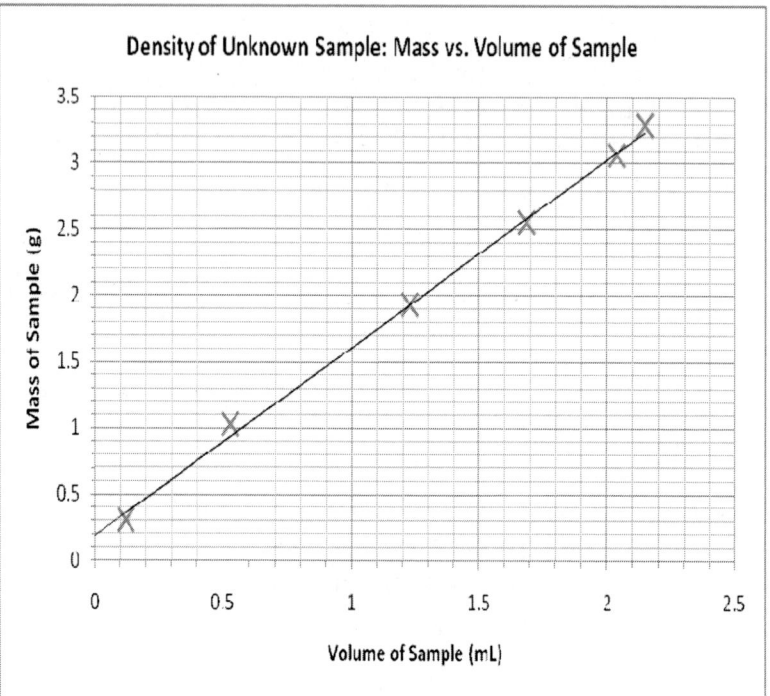

Vol in mL	Mass in g
0.12	0.296
0.53	1.023
1.23	1.928
1.69	2.546
2.04	3.055
2.15	3.281

Density of Unknown Sample: Mass vs. Volume of Sample

d) Using the equation from c) calculate the volume for a sample with a mass of 5.219 g. Show your work.

Answers:

a) Possible points to use for the slope: (0.30 mL, 0.600 g) and (1.90 mL, 2.900 g)
   Note that the x-values (volume) has the same decimal places and units as the data for volume, and the y-values (mass) have the same decimal places and units as the data for mass. The points, by convention, are listed with the x-value first, then the y-value.

$$\text{slope} = \frac{(2.900 \text{ g} - 0.600 \text{ g})}{(1.90 \text{ mL} - 0.30 \text{ mL})} = \frac{2.300 \text{ g}}{1.60 \text{ mL}} = 1.44 \text{ g/mL}$$

b) y-intercept = 0.19 g
   y-intercept is read off the graph where the line intersects with the y-axis at x = 0.0 mL, and recorded to the same decimal places as the y-data.

c) Equation for line is y = (1.44 g/mL)(x) + 0.19 g
   To be more specific, the equation in terms of M and V where M = mass, V = Volume

   $$\boxed{M = (1.44 \text{ g/mL}) V + 0.19 \text{ g}}$$

d) First solve for the unknown (V), then insert 5.219 g for M:

$$V = \frac{(M - 0.19 \text{ g})}{1.44 \text{ g/mL}} = \frac{5.219 \text{ g} - 0.19 \text{ g}}{1.44 \text{ g/mL}} = \frac{5.20\cancel{9} \text{ g}}{1.44 \text{ g/mL}} = 3.51\cancel{6} \text{ mL} = \boxed{3.52 \text{ mL}}$$

Appendix 3: GRAPH PAPER

> **Please do not use other graph paper. Use only what is provided here.** You are to use only a *sharp* pencil in the preparation of your graphs so that if you should make a mistake you can easily correct your mistake. If for some reason you cannot erase your mistake, you can use the backside of the graph paper.

APPENDIX 5: TABLE OF ELECTRONEGATIVITY VALUES

In General Chemistry we use electronegativity values to determine whether a bond is ionic, polar covalent or nonpolar covalent. Most of the time this can be determined just by considering the periodic trends in electronegativity of the elements rather than referring to the table shown below. You are strongly urged to rely on your knowledge of the periodic trends. This table should be used only in the cases when the trends cannot help you make a prediction.

NOTE: This is NOT a Periodic Table!!!

I A	II A		III B	IV B	V B	VI B	VII B		VIII B		I B	II B	III A	IV A	V A	VI A	VII A	VIII A
1 H 2.1																		2 He
3 Li 1.0	4 Be 1.5												5 B 2.0	6 C 2.5	7 N 3.0	8 O 3.5	9 F 4.0	10 Ne
11 Na 0.9	12 Mg 1.2												13 Al 1.5	14 Si 1.8	15 P 2.1	16 S 2.5	17 Cl 3.0	18 Ar
19 K 0.8	20 Ca 1.0		21 Sc 1.3	22 Ti 1.5	23 V 1.6	24 Cr 1.6	25 Mn 1.5	26 Fe 1.8	27 Co 1.8	28 Ni 1.8	29 Cu 1.9	30 Zn 1.6	31 Ga 1.6	32 Ge 1.8	33 As 2.0	34 Se 2.4	35 Br 2.8	36 Kr
37 Rb 0.8	38 Sr 1.0		39 Y 1.2	40 Zr 1.4	41 Nb 1.6	42 Mo 1.8	43 Tc 1.9	44 Ru 2.2	45 Rh 2.2	46 Pd 2.2	47 Ag 1.9	48 Cd 1.7	49 In 1.7	50 Sn 1.8	51 Sb 1.9	52 Te 2.1	53 I 2.5	54 Xe

A-17

Appendix 6: THE FORMAL LAB REPORT

A formal chemistry lab report generally contains specific sections: title page, abstract, introduction, experimental section, data and observations, calculations and results, discussion or results, conclusions, and references. Follow this format meticulously while preparing your formal lab report.

Cover page:
The cover page states general information to help your instructor identify the source of the report. At the top of the page give your name, full name of partner (if any, clearly stated as a partner), name of instructor, course and section number, semester, date experiment was performed and date of submission of report.

Next, state the experiment number, and either copy the title from the lab manual or come up with your own; it should be in the middle of the page and have a bigger font size than the rest of the report. (See sample of a formal lab report in Appendix 7.)

Abstract (Start on new page):
This section is a one-paragraph summary in about 100 to 200 words, in boldface and single spaced. It includes the purpose and importance of the experiment or study, the experimental method used, the major results, and the conclusions drawn. Specific information about the experiment or study must be written in the past tense.

Introduction:
This section provides background information (theory and previous research). It states the goals/objectives. *Include all relevant equations and formulas.* Cite the resources properly. The past tense is used except for well-established facts. A separate paragraph is used for each major point.

Experimental Section:
It consists of two parts: Equipment/Materials and Procedure. The "Procedure" follows Equipment/Materials. The different steps completed during the experiment are described in complete sentences in one or two well-formed paragraphs. Past tense and passive voice are used in this section. Example: *A sample of KHP weighing 0.351g was placed in a 100-mL beaker.*

Data and Observations:
Copy the data table from the laboratory manual. Include all relevant observations in past tense and passive voice.

Calculations and Results:
The Calculations and Results page from the lab manual is to be attached to the back of the report. Show all calculations in details using your data. <u>*Do not describe the steps involved in the calculations*</u>. Label all steps (if appropriate). Calculation setups can be done by hand and attached. Show your results at the end. Scientific notation must be properly expressed:
Acceptable: 2.5×10^{-5} Unacceptable: 2.5E-05   Unacceptable: 2.5*-5

Graphs (full-size paper required), if appropriate, should be included after the conclusions. It should be clearly mentioned that the graph has been attached (after the conclusions). Graphs must be clearly labeled. Give a title for the graph and label the axes. Review Appendix 2.

Discussion of Results:

Discuss your results. Here the results obtained from the experiment are compared with the expected results. If the results are what you expected, indicate which theory or law they are consistent with. If results are inconsistent with expected results, account for the errors. The past tense is used to discuss the experimental findings.

Start this section by giving a statement of your accomplishment. For example, molar concentration of an unknown NaOH solution was determined by performing an acid-base titration.

After that, analyze your results and write one or two short paragraphs to show how each part of your results support or refute what you are trying to prove in the experiment. Depending on the experiment, some discussions will be longer than others. If there were <u>obvious</u> errors, provide possible reasons for the errors and propose changes in the procedure that could help one avoid such errors in the future.

Your instructor will specify which of the post-lab questions are to be included in discussion section of the lab report. In such cases, **DO NOT MAKE ANY SEPARATE LABEL WHILE ANSWERING THE QUESTIONS.**

Conclusion:

Give a <u>short summary</u> of your discussion and conclusion (about one or two sentences). It should not mention anything new that you have not already included in your discussion.

- unacceptable: I have accomplished my objectives. *(too vague)*

- unacceptable: I have learned to use a pipet correctly *(not the purpose of the experiment)*

- acceptable: The density of unknown solid #257 has been determined to be 6.9 g/cm^3, based on 3 trials, with a relative average deviation of 0.07 %. It has been found that the "displacement of water" method based on "Archimedes' principle" works well for determining density of irregularly shaped solid objects.

References*:

In this section, the sources cited in the body of the paper are listed in numerical order. It is a <u>numbered</u> list of sources of information you had cited in the lab report. Note that "references" is different from "bibliography." "Bibliography" is a list of sources of information from which you obtained general knowledge of a particular topic without directly quoting from it. Your formal lab report does not require a "bibliography."

Examples of how references are to be listed based on the American Chemical Society (ACS) Style Guide:

BOOK:
Authors last names, initials, Title (in italics), edition; Publisher: City, State (in 2 letters), Year; pages or chapter

Hamilton, P., Zaman K., Yau, C. *CHEM 122 Experiments in General Chemistry I Laboratory*; Academx: Bel Air, MD, 2013; pp 56-61.

Malone, L. J. *Basic Concepts of Chemistry,* 7th ed.; John Wiley & Sons: St. Louis, MO, 2004, pp 145-148.

HANDBOOK:
CRC Handbook of Chemistry and Physics, 81st ed. Lide, D.R., Ed.; CRC Press: Boca Raton, FL, 2000-2001; p. **4-73**.

The Merck Index: An Encyclopedia of Chemicals, Drugs, and Biologicals, 12th ed. Budavari, S.; O'Neal, M.J.; Smith, A.; Heckelman, P.E.; Kinneary, J.F., Eds.; Merck & Co.: Whitehouse Station, NJ, 1996; entry 4857.

WEBSITES:
http://pubs.acs.org/books/references.shtml (accessed June 17, 2007)
http://chemfinder.cambridgesoft.com (accessed July 14, 2004)

Answers to Post-Lab Questions:

As indicated previously under "Discussion of Results," your instructor will specify certain of the post-lab questions to be included as part of the discussion. Answers to all of the remaining post-lab questions are to be typed and stapled to the back of your lab report. In this section, please include the questions and number them according to how they are listed in the lab manual. The grade will be part of the grade for the discussion section.

Citation*:

Journals published by the American Chemical Society (ACS) utilize two methods of citing references: by superscript numbers, or by italic numbers in parentheses. You may select one method and use it consistently for the entire report. For example, in the body of the report, the literature value is assigned a "reference number" as shown:

The BP is given as 67 °C in the literature.[3]
The BP is given as 67 °C in the literature *(3)*.

The numbering starts with 1 and is continued consecutively throughout the report. If exactly the same reference (including page numbers) is cited, do not use a different reference number.

The author/authors name(s) may be part of the sentence. If two authors are involved, give both names joined by the word "and". If a reference has more than two authors, give only the name of the first author listed followed by "et al." See examples shown below:

The procedure described by Jones[3] was followed.
The procedure described by Jones and Smith[3] was followed.
The procedure described by Hamilton et al.[3] was followed.

*What is the difference between Reference and Citation?

    Citation occurs <u>within</u> the body of the text and a reference number is assigned to each citation.

    Reference is the list of works cited in the text, placed at the end of the document, numbered according to the reference number used in the citation.

Appendix 7:
SAMPLE OF A FORMAL LAB REPORT

Your instructor will provide you with a sample.

Appendix 8: WRITING EXERCISES

Keep in mind the following points in writing a document in science:
- Be clear.
- Be complete.
- Be concise.

This means that you should know exactly what you want to say and do so in as few words as possible, and yet be clear and complete. This is easier said than done and it **takes practice**. For this reason, in addition to the above three points, here is one further point which is not limited to science:

Treat your final document as a first draft and **take time to proofread and revise** for better sentence structure and organization of thought, as well as for correctness in spelling and grammar.

PART I: WRITING IN PASSIVE VOICE

In writing a scientific document, such as a lab report, we tend to write much more in passive voice than in active voice, the reason being that generally <u>who</u> is doing the work, or <u>who</u> is making the observations is not important. Especially when describing a procedure or method, it should be one that anyone skilled in the discipline should be able to do and get similar results. Thus, we do not want to say, "I did this, and I did that." Instead, we say, "This was done, and that was done."

EXAMPLES:

AVOID ACTIVE VOICE	PREFER PASSIVE VOICE
I added 40.0 mL of HCl solution to the Mg strip.	To the Mg strip was added 40.0 mL of HCl solution.
I saw bubbles forming.	Formation of bubbles was observed.

EXERCISE 4.1
Rewrite the following sentences in passive voice.

1. I transferred the contents of the beaker to the graduated cylinder.
2. I found the slope of my graph is 0.04 g/min.
3. I added 3 drops of bromophenol indicator to the Erlenmeyer flask.

PART II: WHEN IS PASSIVE VOICE NOT NECESSARY?

Note that we are not saying passive voice is always necessary. It is generally used when the doer is unknown or unimportant, as discussed in Part I. It is also used occasionally to change the tone for variety. As a rule of thumb, if the first person singular pronoun is not involved, the active voice is preferred because it is less wordy, more direct and at times, less awkward.

Using first person plural pronouns, such as "we" is acceptable when "we" refer to an inclusive "I" and the readers. This is a way to indicate that not just the writer but everyone reading the document is considered. If necessary, use "one" instead of "I" or "you." See examples that follow.

EXAMPLES OF ACCEPTABLE USE OF "WE" OR "ONE"

> From the color of the endpoint we can conclude that the equivalence point has been reached.

> From the color of the endpoint one can conclude that the equivalence point has been reached.

> We can conclude definitively that chloride ions are present in the unknown.

> One can conclude definitively that chloride ions are present in the unknown.

PART III: WRITING EXPERIMENTAL PROCEDURES OR METHODS

When writing a lab report, the experiment has already been performed. The "experimental methods" or "experimental procedure" is meant to describe what you had ALREADY done. You are not writing a lab manual to tell others what they have to do, or what you are about to do. Thus, <u>past tense</u> rather than present tense is to be used, in addition to writing in passive voice.

EXAMPLE:

INCORRECT	PREFER PAST TENSE, PASSIVE VOICE
Weigh and record the flask containing the residue.	The flask containing the residue was weighed and recorded.
I heated the yarn in a clean test tube.	The yarn was heated in a clean test tube.

EXERCISE 4.2:
Rewrite the following sentences in the preferred tense and voice (past tense, passive voice).
4. Cover the calorimeter with the cardboard lid and insert a second temperature probe through one of the holes in the lid.
5. Stir the mixture thoroughly with a clean spoonula and grind any lumps that you may find into fine grains.
6. Examine the popped corn and record the number of duds.

APPENDIX 8: WRITING EXERCISES   A-31

PART IV: WRITING THE EXPERIMENTAL METHOD IN AN ABSTRACT

In an abstract, the "experimental method" must not include details of size of equipment and quantities of samples used unless this information is critical to the method (which usually is not the case). What is to be included in an abstract should ***not*** be the "procedure" but just a "method" used. See example that follows in the difference between "procedure" and "method."

EXAMPLE OF A PROCEDURE:
Record the unknown number. Place the unknown metal on the balance pan and record the mass to 3 decimal places. Next place exactly 5.00 mL of water in a 10-mL graduated cylinder. Slower slide the unknown metal into the grad cylinder, taking care not to let any water splash out. Record the total volume to 2 decimal places. Calculate the volume of the metal by subtracting 5.00 mL from the total volume. Calculate the density of the metal by dividing the mass of the metal by the volume of the metal.

EXAMPLE OF AN EXPERIMENTAL METHOD (AS IN AN ABSTRACT):
The volume of the unknown metal was determined by its displacement of water and the mass was determined on a balance. From the volume and mass thus obtained, the density of the metal was calculated.

EXERCISE 4.3:
7. Write the experiment method for an abstract from the procedure described below:

> **Effect of Eye-Level on Accuracy of Reading Volumes**
>
> 1. Place **exactly** 7.00 mL of deionized water into a 10-mL grad cylinder. Use a disposable pipet to help you add or remove excess water so that the bottom of the meniscus is at exactly the 7-mL mark when held at eye-level.
> 2. Hold the cylinder so that the meniscus is well above your eye-level. Record the volume. (Remember to record to the correct sig. fig.)
> 3. Repeat with the cylinder at eye-level and below eye-level.
> 4. Complete the calculations specified on the Calculations & Results page (p. 48).

Appendix 9: HOW TO ENTER SUBSCRIPTS & SUPERSCRIPTS

In chemistry documents you will often need to include the formula of chemicals that requires the use of subscripts and superscripts. For example, entering "MgSO4" is totally unacceptable as it should be "$MgSO_4$." Your instructor may allow you to hand write in the subscript "4" but you will probably need to write it in numerous times within the same document and you will likely miss some of them. It is highly recommended that you learn to enter subscripts and superscripts directly on the computer.

PROCEDURE FOR WORD 2010:
Most CCBC college computers are using Word 2010. The following procedure is specifically for this version of Word. The procedure for older versions of Word is on the next page.

To enter a number as a subscript, click on the **Home** tab at the top left corner, and the x_2 in the **Font** area. The superscript mode is accessed in a similar manner.

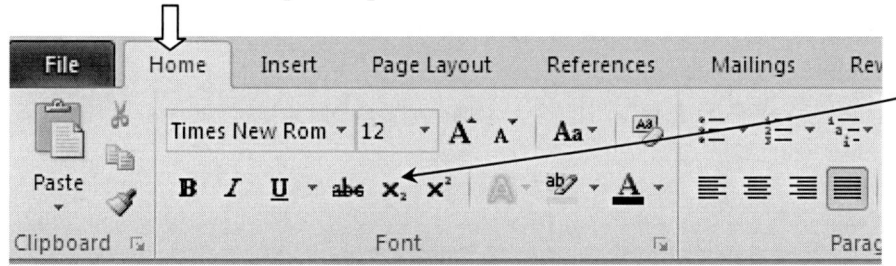

As an example, follow these steps to enter $MgSO_4$:
1. Type "MgSO" and click on the **Home** tab at the top left corner, and the x_2 in the **Font** area.
2. Type "4", and click on x_2 again. This toggles your keyboard back to standard mode.

As an example, follow these steps to enter "SO_4^{2-}"
1. Type "SO" and click on the **Home** tab, and on x_2, then type "4."
2. Click on x^2, type "2-" and then on x^2 to toggle back to the standard mode.

Alternatively you can use the keyboard shortcut:
Ctrl and = (both keys together) put you in the subscript mode.
Ctrl, **Shift** and = (all three keys together) put you in the superscript mode.
Pressing this combination of keys will toggle you back to the regular mode.

Now test yourself:

Open a Word 2010 blank document and try typing the following with the subscripts and superscripts properly placed:
H_2CO_3 and CO_3^{2-}

A-34   APPENDIX 9: HOW TO ENTER SUBSCRIPTS & SUPERSCRIPTS

PROCEDURE FOR WORD 97-2003
First check to see whether the icons for x_2 and x^2 icons are on your toolbar at the top. A previous user may already have placed them there. If not, follow these directions to place them on your toolbar.

TO PLACE THE x_2 AND x^2 ICONS ON YOUR TOOLBAR:
1. Click on **Tools** and then on **Customize**.
2. At the top you will see three tabs: **Toolbars, Commands,** and **Options.** Select **Commands**.
3. You should now see two lists: **Categories** and **Commands.** In the **Categories** list (on the left), select **Format**.
4. In the **Commands** list (on the right), scroll down until you find x_2 **Subscript** and drag it up to your toolbar. If you are using your own personal computer (as opposed to a public computer), this icon should remain on your toolbar from now on.
5. In the **Commands** list, find x^2 **Superscript** and drag it up to your toolbar.

TO USE THE x_2 AND x^2 ICONS THAT YOU PLACED ON YOUR TOOLBAR:
To enter a number as a subscript, click on the x_2 icon, type the number, then click on the x_2 icon again to toggle it back to the standard mode. The superscript mode is accessed in a similar manner.

ALTERNATIVELY YOU CAN USE THE KEYBOARD SHORTCUT:
Ctrl and = (both keys together) put you in the subscript mode.
Ctrl, Shift and = (all three keys together) put you in the superscript mode.
Pressing this combination of keys will toggle you back to the regular mode.

As an example, to enter $MgSO_4$:
   Type "MgSO", click on x_2, type "4", then click on x_2 again.
Or use keyboard shortcut:
   Type "MgSO", press **Ctrl** and = together, type "4", press **Ctrl** and = again to toggle back.
As an example, to enter SO_4^{2-} :
   Type "SO", click on x_2, type "4", click on x^2, type "2-", then click on x^2 again.
Or use keyboard shortcut:
   Type "SO", press **Ctrl** and = together, type "4".
   Press **Ctrl, Shift** and = all three keys together, type "2-". Press **Ctrl** and **Shift** and = again to toggle back.

Now test yourself:
Try typing the following with the subscripts and superscripts properly placed:
      H_2CO_3 and CO_3^{2-}